First Published, 2000
Reprinted, 2018

ISBN: 978-81-87498-02-5 (Harbound)

Publisher's note:
Every possible effort has been made to ensure that the information contained in this book is accurate at the time of going to press, and the publisher and author cannot accept responsibility for any errors or omissions, however caused. No responsibility for loss or damage occasioned to any person acting, or refraining from action, as a result of the material in this publication can be accepted by the editor, the publisher or the author. The Publisher is not associated with any product or vendor mentioned in the book. The contents of this work are intended to further general scientific research, understanding and discussion only. Readers should consult with a specialist where appropriate.
Every effort has been made to trace the owners of copyright material used in this book, if any. The author and the publisher will be grateful for any omission brought to their notice for acknowledgement in the future editions of the book.

Published by : **Regency Publications**
A Division of
Astral International Pvt. Ltd.
– ISO 9001:2008 Certified Company –
4736/23, Ansari Road, Darya Ganj
New Delhi - 110 002
Phone: 011-4354 9197, 2327 8134
E-mail: info@astralint.com
Website: www.astralint.com

PREFACE

To feed the growing population there is an urgent need to increase the production of food grains. Grasses attracted the attention of plant breeders because most of the food grains belong to the family Poaceae (Gramineae). The family also attracted the attention of plant breeders because of the occurrence of Apomixis. In recent years, hybridization between sexual crops and their closest apomictic wild relatives have been performed with a view to obtain homozygous lines, fixing heterosis and getting non-segregation population from hybrid with unique combination of beneficial characters from the parent forms.

There is no comprehensive account on Embryology and apomixis in such an important family. To fill this gap this book is written.

Several botanists helped us by sending reprints of their papers and sharing their literature. These include Dr. B.L. Burson of Texas A&M University, Dr. J.G. Carman of Utah State University, Dr. Ozias-Akins of University of Georgia, Dr. M.D. Peel of North Dakota State University, Dr. W. Hanna of USDA Coastal Plain Experiment Station, Dr. O. Leblanc of CIMMYT, Mexico, Dr. J.J. Spies of University of Free State, Dr. T. Verboom of University of Cape Town, Dr. C.L. Quarin of Universidad Nacional del Nordeste, Prof. E. Battaglia from Italy, Dr. R.K. Bhanwra from Punjab University and Prof. A.N. Sindhe of Mysore University. Authors are grateful to all these scientists.

This is only an attempt to summarise the work. We don't claim this as a perfect account. Had we tried for perfection we could have never completed this work. We request the readers of this book to bring to our notice errors and omissions.

CONTENTS

CHAPTER 1

INTRODUCTION

Poaceae *(nom. alter.* Gramineae) is the third largest family of flowering plants. According to Tzvelev (1989) the family comprises of 899 genera and 10,300 species. They are distributed in various climates, soils and elevations on the surface of the globe. The family plays a vital role both in man's economic activity and in the composition of natural plant communities.

Morphologically the family is characterised by compound spike inflorescence, bisexual or unisexual zygomorphic flowers subtended by several distichous imbricate glumes, perianth represented by 2–3 minute hypogynous scaly lodicules or sometimes absent, stamens three to six, ovary superior unilocular with single ovule and fruit caryopsis. Except bamboos majority of the grasses are herbaceous annual or perennial plants.

Many species are the important source of nutritious herbage for livestock. Cereals such as wheat, rice and maize are the most important food plants of mankind. Members of many other grass genera are used to make flour and meal. Sugar cane which is widely cultivated in the tropics, is one of the major source of such a valuable product as sugar. Many bamboos and some other tall grasses make good construction materials, as well as providing raw material for the production of paper and various other things.

The study of grass embryology has two aspects of vital importance (Bhanwra, 1988). Firstly the phenomena of apomixis is of frequent occurrence (Brown and Emery, 1958). Secondly the embryological characters have been utilized in deducing inter-relationships between subfamilies (Chandra, 1963b; Venkateswarlu and Devi, 1964; Reeder, 1957; Reddy, 1977), tribes (Maze and Bohm, 1977), genera (Maze and Bohm, 1973; Maze *et al.*, 1970, 1971; Kam and Maze, 1974) and even species (Aulback-Smith and Herr, 1984).

The study of angiosperm embryology has a special claim to attention in connection with the breeding and cultivation of crop and pasture plants. In recent years, hybridization between sexual crops and their closest apomictic wild relatives have been performed with a view to obtain homozygous lines, fixing heterosis and getting non-segregation population from hybrid with unique combination of beneficial characters from the parent forms (Asker, 1979).

The increasing usage of grasses, reintroduction of new food plants into culture and the necessity of selection in cultivated species make the field of embyrology of this family exceptionally important. The number of scientific works relating to grass embryology increasing every year due to the occurrence of the phenomenon of apomixis. Extensive embryological work has been carried out in this family by several embryologists. Inspite of this extensive study adequate information is not available for several members and several species remain uninvestigated. Inadequate investigation may be due to the various technical difficulties while processing the material or sometimes even the structure of the inflorescence or floret itself poses problem.

CHAPTER 2

MICROSPORANGIUM, MICROSPOROGENESIS AND MALE GAMETOPHYTE

The anther in all the members of Poaceae is tetrasporangiate. However in *Bromus unioloides* (Bhanwra and Choda, 1980; Bhanwra, 1988) bisporangiate anthers have been reported.

The stamen primordia arise centripetally as rounded or crescent-shaped projections on the floral receptacle. These primordia comprise dermatogen covering the multicellular hump-like tissue. In *Zea mays* sequential nuclear divisions and fusions are correlated with the differentiation of microsporangium (Korobova, 1971).

The young anther comprises of an oval shaped mass of meristematic cells surrounded by the epidermis. During the further development, the anther attains four-lobed contour. In each of the four anther lobes a row of hypodermal cells become differentiated by their larger size, radial elongation, dense cytoplasm and more conspicuous nuclei (Fig. 1A). These form the male archesporium. The extent of these archesporial cells varies both length wise and breadth wise. The male archesporium is usually single-layered (Fig. 2A). However two-layered male archesporium is met with in *Pennisetum typhoideum* (Narayanaswami, 1953) and two- or three-layered archesporium is reported in *Dendrocalamus hamiltonii* (Harigopal and Manasi Ram, 1981).

The archesporial cell(s) expand radially and undergo periclinal division resulting in the formation of primary parietal layer towards the epidermis and primary sporogenous layer towards the interior of the anther (Figs. 1 B,C, 2B,C). Increase in the size of primary sporogenous cells and anticlinal divisions of the parietal ones leads to a considerable size difference in these two types of cells. The cells of the primary parietal layer undergoe one more periclinal division giving rise to the formation of outer parietal layer and inner parietal layer (Figs. 1D, 2C–G). The outer parietal layer develops fibrous thickenings at maturity and functions as fibrous endothecium. The inner parietal layer undegoes one more periclinal division resulting in the formation of middle layer and tapetal layer (Figs. 1E, 2H–J). At a stage when the anther has attained the deeply four-lobed contour the wall layers are discernible as tapetum, middle layer and tapetum (Fig. 3A–D). The schematic representtation of the anther wall development is as follows:

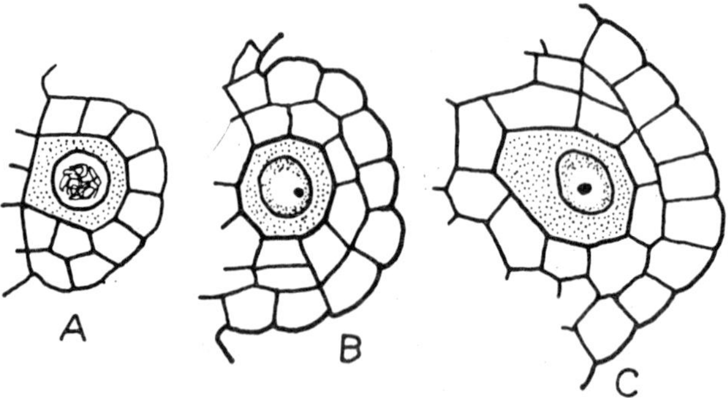

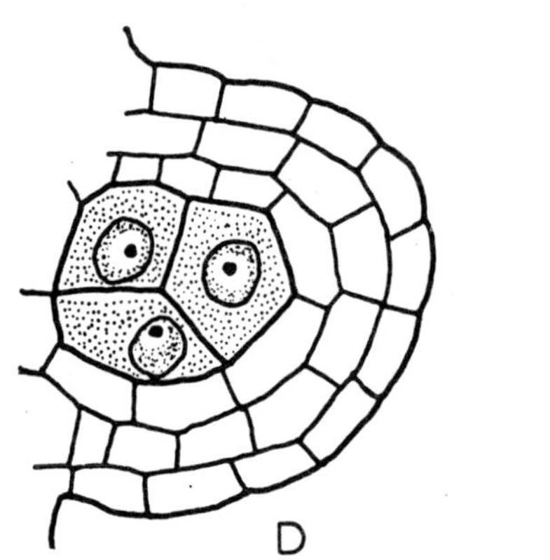

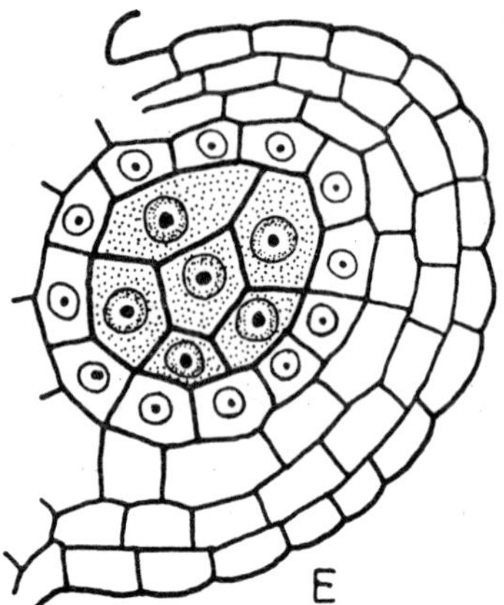

Fig. 1. A. T.s. of anther lobe showing male archesporium. B, C. T.s. anther lobes showing sporogenous cell and parietal layer. D. T.s. of anther lobe showing sporogenous cells and two parietal layers. E. T.s. of anther lobe showing pollen mother cells and anther wall. A, B, E. *Brachiaria reptans* (Febulaus & Pullaiah, 1991 b). C. *Eragrostis bifaria* (Febulaus & Pullaiah, 1992 b). D. *Panicum repens* (Febulaus & Pullaiah 1992 a).

This type anther wall development, according to Davis (1966), corresponds to the Monocotyledonous type. Thus the anther wall comprises of four layers namely epidermis, endothecium, middle layer and tapetum.

The single-layered epidermis may remain intact (Davis, 1966). During the course of maturation of the anther the epidermal cells undergo repeated anticlinal divisions in order to cope up with the rapidly enlarging internal tissue. In mature anther these cells expand lengthwise. The epidermal cells get stretched both tangentially and longitudinally at maturity. In *Triticale* (Bhandari and

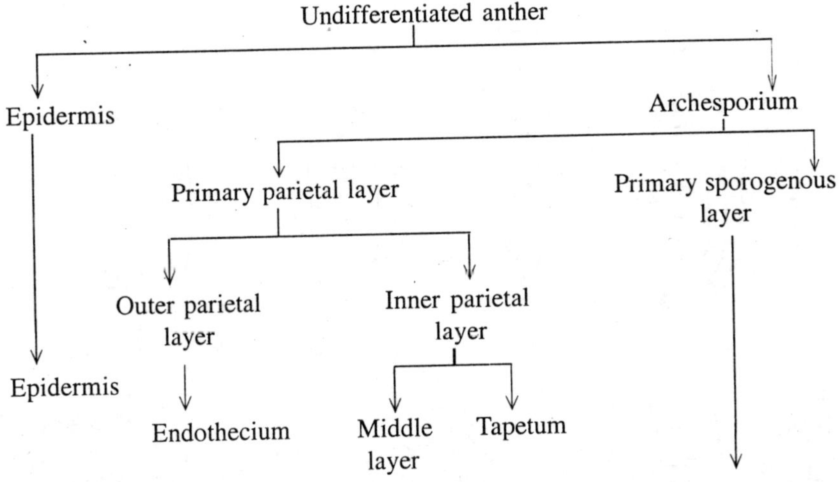

Khosla, 1988) the epidermal cells develop cuticular fibrillar projections. The cuticle in corn comprises homogeneous and reticular layers composed of alkanes (Cheng *et al.*, 1981).

The endothecial cells develop characterstic fibrous thickenings at the time of pollen mitosis. These arise along the inner tangential walls, extend outwards and upwards and ending at outer tangential walls of each cell. They attain their maximum development when the anther is ready to dehisce for the the discharge of pollen. In Poaceae endothecium has been studied by Manning and Linder (1990). According to them two types of thickenings are present. In Type I, which is characteristic of most genera, the cells are rectangular, square in radial view, with the long axis normal to the anther axis. Cells at stomium Ca. isodiametric and at the connective 2-3 rows of rectangular cells are oriented parallel to the anther axis. Thickenings fully developed in cells in the outer half of the anther thecae, consisting of a complete base plate with strips extending up to the anticlinal walls and shortly overtopping into the outer periclinal walls; moving centripetally from the connective the cell thickenings commence with thickened pegs on the anticlinal walls and these develop a thin, complete base plate. At the connective, cells have a thickened tympanic/radial base plate and overtopping anticlinal strips (Manning and Linder, 1990).

In the genera *Merxmuellera, Pentaschistis* and *Tribolium* the cells are longer and more fusiform and some of the anticlinal strips are continuous across the outer periclinal wall. In *Sporobolus* and *Eragrostis* the cells are weakly thickened except in the outer 1/4 of the thecae and anticlinal thickenings are often continuous across the outer periclinal walls. In *Arundo* the cells are fusiform. In *Zea* and *Coix* thickened cells are mostly restricted to the apex of the anther, but with some at the base (Manning and Linder, 1990).

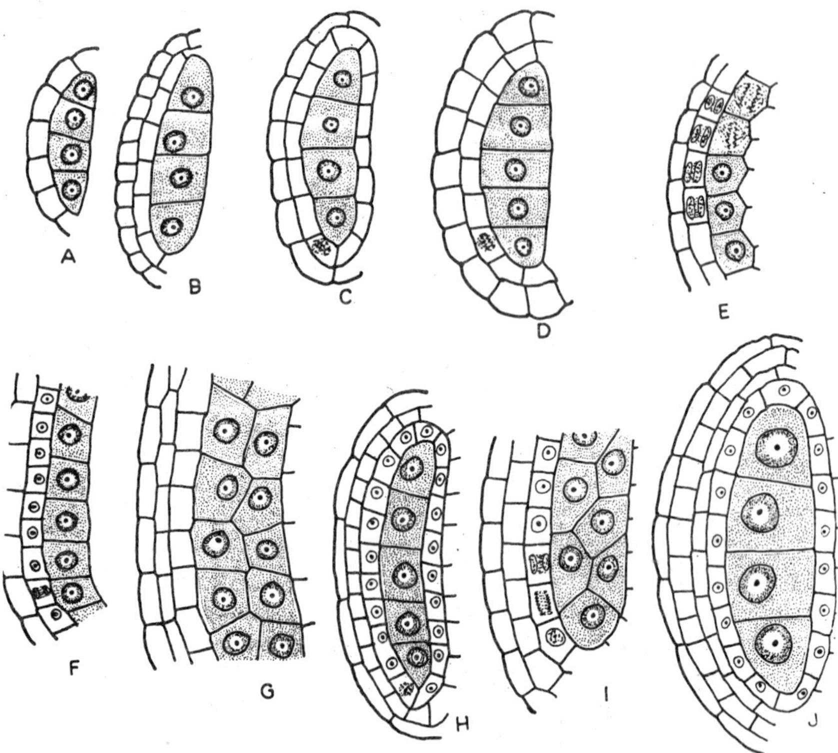

Fig. 2. L.s. of anther lobes showing A. Epidermis and male archesporium. B. Primary parietal layer and sporogenous cells. C-F. Divisions in primary parietal layer and sporogenous cells. G, H. Two parietal layers and sporogenous cells. I. Division of inner parietal layer and sporogenous cells. J. Anther wall and pollen mother. A, B. *Melanocenchris jacquemontii* (Febulaus & Pullaiah, 1997); C, H, J. *Eragrostis viscosa* (Febulaus & Pullaiah, 1990). D. *Chloris roxburghiana* (Febulaus & Pullaiah, 1991a). E. *Brachiaria eruciformis* (Febulaus, 1991). F, I. *Eragrostiella bifaria* (Febulaus & Pullaiah, 1992 b). G. *Panicumrepens* (Febulaus & Pullaiah, 1992a).

In Type II which occurs in genera like *Arundinaria, Bambusa, Leersia* and *Olyra* base plates are absent and the thickenings are basically annular. In *Oryza* the thickenings are 'U' shaped. In *Bambusa* the cells are often rather irregularly shaped and the thickening in consequence have some basal branching. In *Leersia* the thickenings on the anticlinal walls are heavy while those on the periclinal walls are tenuous or nearly absent, resulting in two opposing C-shaped thickenings.

Juliano and Aldama (1937) reported that in *Oryza* one or two perietal layers become crushed and disorganised during the development of anther so that a fibrous layer is absent and the epidermis abuts directly on the tapetum. In *Dendrocalamus hamiltonii* the endothecial cells are gorged with compound

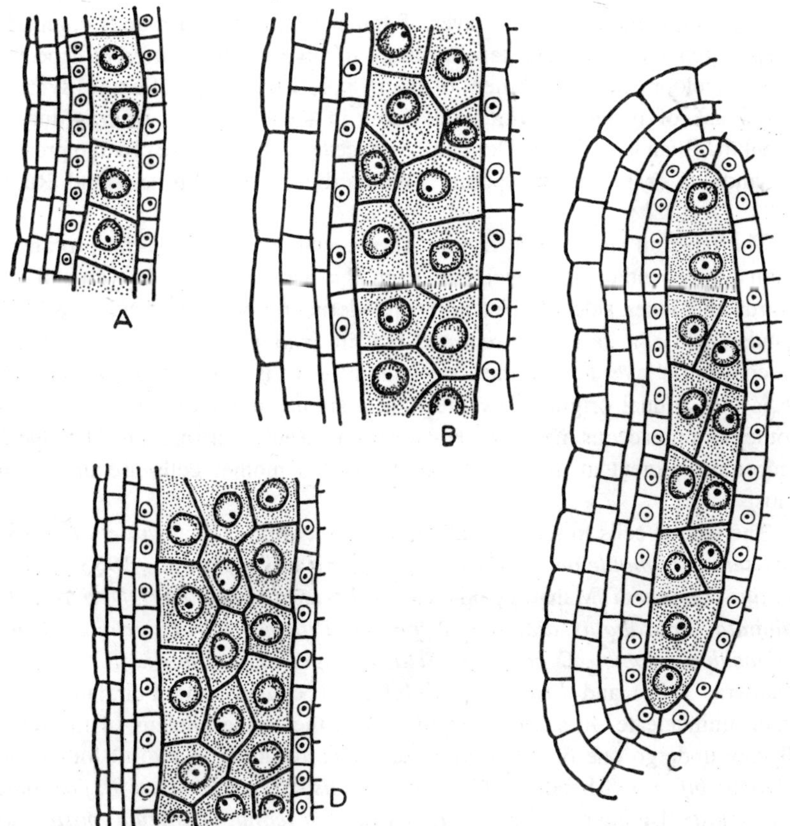

Fig. 3. A–D. L.s. part of anther lobes showing epidermis, endothecium, middle layer and anther tapetum. A. *Aristida hystrix* (Febulaus & Pullaiah, 1993b). B. *Cymbopogon caesius* (Febulaus, 1991). C. *Panicum notatum* (Febulaus, 1992). D. *Branchiaria reptans* (Febulaus & Pullaiah, 1991b)

starch grains (Harigopal and Manasi Ram, 1981). In *Spinifex littoreus* Lakshmanan and Jayalakshmi (1980) reported that fibrous thickening are absent in endothecial cells.

The endothecium is usually limited to the protuberent part of the microsporangium. It originates from the parietal layer. Bhandari and Koshla (1982) reported that in *Triticale* a complete ring of endothecial layer is surrounding the tapetum and sporogenous tissue. This originates exclusively from the parietal layer.

The presence of fibrous thickenings, differential expansion of the outer and inner tangential walls and the hygroscopic nature of these endothecial cells helps in the dehiscence of anther at maturity.

The middle layer in all the members of the family is ephemeral. The cells become compressed and degenerated before the microsporangium is fully matured to discharge pollen.

Tapetum is the inner most layer of anther wall and it surrounds completely the sporogenous tissue. The tapetum may involve in three different aspects of pollen development: nourishment of microspores, formation of exine, synthesis and release of materials which take part in the deposition of tryphine and pollenkitt (Bhandari, 1984). The family Poaceae is characterised by the presence of glandular or secretary tapetum (Wunderlich, 1954, Furness and Rudall, 1998).

The tapetum shows dual origin. The tapetal cells towards the protuberant region of the anther are developed from the parietal layer. The tapetal cells towards the inner side of the anther are derived from the connective tissue. Bhandari and Khosla (1982) reported that in *Triticale* the tapetum originating as a concentric layer around the sporogenous tissue is solely from parietal tissue and is not of dual origin. They presume a similar mode of origin of tapetum in other Graminaceous members. However this needs confirmation. The tapetum is most promiment at a stage when the pollen mother cells are undergoing meiosis.

The tapetal cells in some members like *Poa annua* (Bhanwra *et al.*., 1985), *Pennisetum typhoideum, Panicum miliare* (Narayanaswami, 1953, 1955a), *Eleusine compressa* (Mahalingappa, 1977), *Thyrsostachys oliveri, Desmostachya bipinnata, Sporobolus diander, Tripogon filiformis, Brachiaria distachya, Echinochloa colonum, E. crusgalli, Setaria glauca, S. intermedia, S. verticillata* (Bhanwra, 1988) and *Eragrostiella bifaria* (Febulaus and Pullaiah, 1992b) remain uninucleate. In some other members, the single nucleus of the tapetal cell may undergo one or two mitotic divisions resulting in two to four nuclei. In *Tragus biflorus* (Chandra, 1975a), *Eragrostis ciliaris, E. nigra, E. poaeoides, E. tremula* (Bhanwra, 1986a), *Eleusine coracana, Panicum miliaceum, Echinochloa frumentacea* (Narayanswami, 1952, 1955a,b), *Eleusine indica, Dactyloctenium aegyptium* (Chandra, 1976), *Cymbopogon caesius* and *Dicathium pseudoischaemum*, they become binucleate. Trinucleate tapetal cells have been observed in *Brachiaria eruciformis, B. reptans, Aristida funiculata, Eragrostis viscosa* while four nucleate tapetal cells have been observed in *Aristida mutabilis* and *Perotis indica* (Febulaus and Pullaiah, 1991a,b; 1992a,b; 1993a,b; 1995; 1996). In *Zea mays* (Bianchi, 1959; Carniel, 1961) only one mitosis occurs in tapetal cells; it produces binucleate (2n + 2n) or uninucleate (4n) cells depending on whether the mitotic process is normal or restitutional.

Ultrastructure of glandular tapetum has been studied in some taxa such as *Poa annua* (Rowley, 1963), *Sorghum bicolor* (Christensen *et al.,* 1972) and *Avena sativa* (Steer, 1977). In *Avena* (Steer, 1977) newly formed tapetal cells show plasmodesmatal connections between the adjacent cells and also between the sporogenous cells. Microtubules run parallel to the long axis of anther along the tangential walls and tangentially or radially along the radial walls. Fewer ribosomes are reported in the tapetal cells. The wall of the tapetal cells in *Avena* is thin and comprise the middle lamella and a small amount of cellulosic primary wall composed of numerous fine, lightly stained fibrils.

Christensen *et al.*, (1972) reported that in *Sorghum* the tapetal cell cytoplasm contains distinct dictyosomes, vesicles, mitochondria and an extensive and enlarged membrane bound tubular system of endoplasmic recticulum. Granular bodies called as Ubisch bodies, also known as Orbicules, have been recorded along the tangential surface of anther tapetal cells facing anther locule. Chemically, these are composed of sporopollenin. El-Ghazaly and Jensen (1986) studied orbicules in *Triticum aestivum* and said that tapetal endoplasmic reticulum is the site of origin of orbicules. Orbicules represent a transport mechanism for sporopollenin between the tapetum and the developing microspores and thus take active part in sporoderm formation. The idea that orbicules might contribute to the development of exine was first raised by Maheshwari (1950). Protoplasmic strands were observed in *Poa annuna* by Rowley (1962) extending from the tapetum to the spinules on the pollen exine. Banerjee and Barghoorn (1971) showed in some grasses that spinules of orbicules are in contact with those on the pollen exine through strands of sporopollenin, indicating a track for the possible transfer of substrate from the tapetal cells to the .exine. In *Triticum*, microfilaments are observed occasionally connecting orbicule spinules with the microspore wall (El-Ghazaly and Jensen, 1986). Banerjee and Barghoorn (1971) found in some grasses, immediately prior to anther dehiscence, some depolymerisation of the mature sporopollenin, "possibly caused by secretion from the centrum of the Ubisch bodies." In the same study it is observed that a large number of new spinules are produced at pollen maturity, apparently by the orbicules. This would suggest an important role for orbicules in sporoderm formation in grasses.

In grasses orbicules have been reported in *Avena sativa* (Steer, 1977; Ogorodnikova, 1986), *Dactylis glomerata* (Banerjee, 1967; Pacini, 1990), *Eleusine africana, E. compressa* (Mahalingappa, 1977), *Lolium perenne* (Pacini *et al.*, 1992), *Oryza sativa* (Nishiyama, 1970, 1976), *Phleum pratense* (Banerjee, 1967; Rowley and Skvarla, 1974), *Poa annuna* (Rowley *et al.*, 1959; Rowley, 1962b, 1963), *Saccharum spontaneum* (Roques and Feldman, 1996), *Secale cereale* (Ogordnikova, 1986), *Sorghum bicolor* (Christensen *et al.*, 1972), *Spartina patens* (Banerjee, 1967), *Triticum aestivum* (El-Ghazaly and Jensen, 1985, 1986b, 1987) and *Zea mays* (Banerjee, 1967; Skvarla and Larson, 1966). According to Banerjee (1967) the tapetal membrane in grasses consists of three layers. (i) Fenestrated layer is present immediately outside the tapetal protoplast, (ii) reticular layer and (iii) the Ubisch granules layer. However Christensen *et al.* (1972) reported only two layers in *Sorghum*. According to them the reticulate layer is absent. Untawale *et al.* (1969) reported Ubisch granules in *Eragrostis unioloides*. The tapetum in *Cynodon dactylon* (Mahalingappa, 1978) remain persistent as a thin band adhering to the inner wall of the endothecium. These cells develop ubisch bodies on this tangential walls.

Studies on cytoplasmic male sterile (CMS) lines indicate that tapetum initiate the process of absorbtion of pollen through altered physical or physiological factors. In *Triticum* (Joppa *et al.*, 1966) and *Sorghum* (Alam and Sandal, 1967) the poorly developed vasculature in the stamens of CMS lines blocked the transport of nutrients and so there was inadequate accumulation of starch on the anther tapetum of CMS plants. This will lead to initiate pollen abortion. Raj (1969) reported that in *Sorghum* the tapetal cells enlarge considerably and occlude the entire pollen sac. Due to this the sporogonous tissue is squeezed and crushed.

Young *et al.* (1979) reported that in male sterile hybrid the pollen-wall malformations were very frequent in meiotically unstable CMS lines. The unusual feature reported by them is the formation of orbicular wall completely surrounding each tapetal cell. In male sterile *Sorghum* the tapetal cells undergo unequal radial elongation. The radial walls of these tapetal cells break down and their protoplast coalasce to form a syncitium. In some anthers the tapetal cells become highly vacuolate and their walls excessively thickened (Narayana, 1976).

Apart from tapetum, endothecium or conducting strand of the filament may also influence the abortive process. Reddy and Reddi (1974) consider that eventhough the tapetal layer is persistent, the other abnormalities are operative in some CMS lines. They observed a positive correlation between pollen sterility and thickness of endothecium at maturity, consequently, the anthers are non-dehiscent and prevent the release of pollen.

The wall of the mature microsporangium consists of epidermis and fibrous endothecium. In the mature anther the adjacent lobes become confluent by the degeneration of partition walls at 3-celled pollen grain stage.

MICROSPOROGENESIS

The primary sprogenous cells may directly function as pollen mother cells or they undergo only few transverse divisions and single vertical or oblique division resulting in two rows of pollen mother cells or they may undergo a few mitotic divisions resulting in a moderate mass of pollen mother cells. In *Perotis hordeiformis* (Venkateswarlu and Devi, 1964) single axial row of sprogenous cells are reported in anther lobe. The number of sporogenous cells in each anther ranges from five to about eighty-five. It ranges from 56 to 85 in bambusoid grasses, 5–36 in chloroid grasses and 8–44 in panicoid grasses (Bhanwra, 1988).

Narayana *et al.* (1976) reported cytoplasmic strands between some or all the spore mother cells in a sporangium of male sterile *Sorghum*. In some instances he reported migration of nuclei at different stages of meiosis and migration of cytoplasm along with nuclei from one spore mother cell to another. Such process is known as cytomixis. This is also reported in some species of *Cymbopogon* mainly due to cytological abnormalities (Chandra, 1976).

Kamra (1960) for the first time reported the occurrence of bi- and multinucleate pollen mother cells in barley. Bi-nucleate pollen mother cells are more frequent (2.4%) as compared to tri- or quadri-nucleate ones (0.1% each). This was true for both normal and mutant varieties of barley. Such phenomenon may be attributed to failure of wall formation during premeiotic mitosis in sporogenous cells. Kempanna *et al.* (1971) reported that out of the 82 lines of Triticales, polynucleate pollen mother cells have been observed in one line. These pollen mother cells contain 2–6 nuclei.

The pollen mother cells become rounded off and undergo meiosis (Figs. 4A–X, 5A–Q). The meiosis in pollen mother cells is normal and cytokinesis is of successive type (Fig. 4 E,F, M,N,W,X) in all sexually reproducing members of the family Poaceae (Artschwager *et al.,* 1929; Chandra, 1963a; Gould, 1970; Christensen *et al.,* 1972; Narayanaswami, 1952, 1953, 1954, 1955a,b,c, 1956; Muniyamma 1978; Bhanwra *et al.,* 1981b; 1982; Bhanwra, 1988; Febulaus and Pullaiah, 1991a,b; 1992a,b; 1993a,b; 1995, 1996). Though successive cytokinesis is of common occurrence in the family Poaceae Artschwager and McGuire (1949) reported simultaneous cytokinesis in *Sorghum vulgare*. In *Themeda australis* Woodland (1964) also reported such simultaneous cytokinesis. From the occasional occurrence of tetrahedral microspore tetrads in *Brachiaria reptans, Cenchrus ciliaris* and *Cymbopogon caesius* it can be assumed that simultaneous cytokinesis also occurs occasionally in these species along with the normal successive cytokinesis (Febulaus and Pullaiah, 1991b, 1995). In *Spinifex littoreus* both simultaneous and successive cytokinesis occur in microspre mother cells, however occurrence of successive cytokinesis is more frequent (Jayalakshmi and Lakshmanan, 1982).

The microspore tetrads in the Poaceae are usually isobilateral (Figs. 1F,N,X, 2P), but sometimes decussate and very rarely tetrahedral. Linear, T-shaped and decussate tetrads or oblique wall formation in one or both the dyad cells besides the isobilateral tetrads have been reported in *Eleusine coracana* (Narayanaswami, 1952), *Echinochloa frumentacea* (Narayanaswami, 1955b), *E. stagnina* (Muniyamma, 1978), *Setaria verticillata* (Bhanwra *et al.,* 1980), *Hordeum hexastichen* (Vegl, 1947), *Stipa ischu* (Gould, 1970) and *Zizania aquatica* (Weir and Dale, 1960). In *Panicum miliaceum* Moskova and Mandlikova (1970) reported isobilateral, linear and T-shaped microspore tetrads in addition to normal isobilateral tetrads. In *Cenchrus biflorus* and *Echinochloa crusgalli* linear microspre tetrads are reported alongwith the isobilateral ones. the formation of linear tetrads is however very rare (Bhanwra, 1986). In *Saccharum benghalense* in addition to the isobilateral microspore tetrads decussate tetrads are also reported (Bhanwra, 1988). Basavaiah and Murthy (1989) noticed the occurrence of rarely T-shaped and linear microspore tetrads in *Urochloa panicoides*. In *Spinifex littoreus* decussate, tetrahedral and linear microspore tetrads have been recorded in addition to the isobilateral microspore tetrads

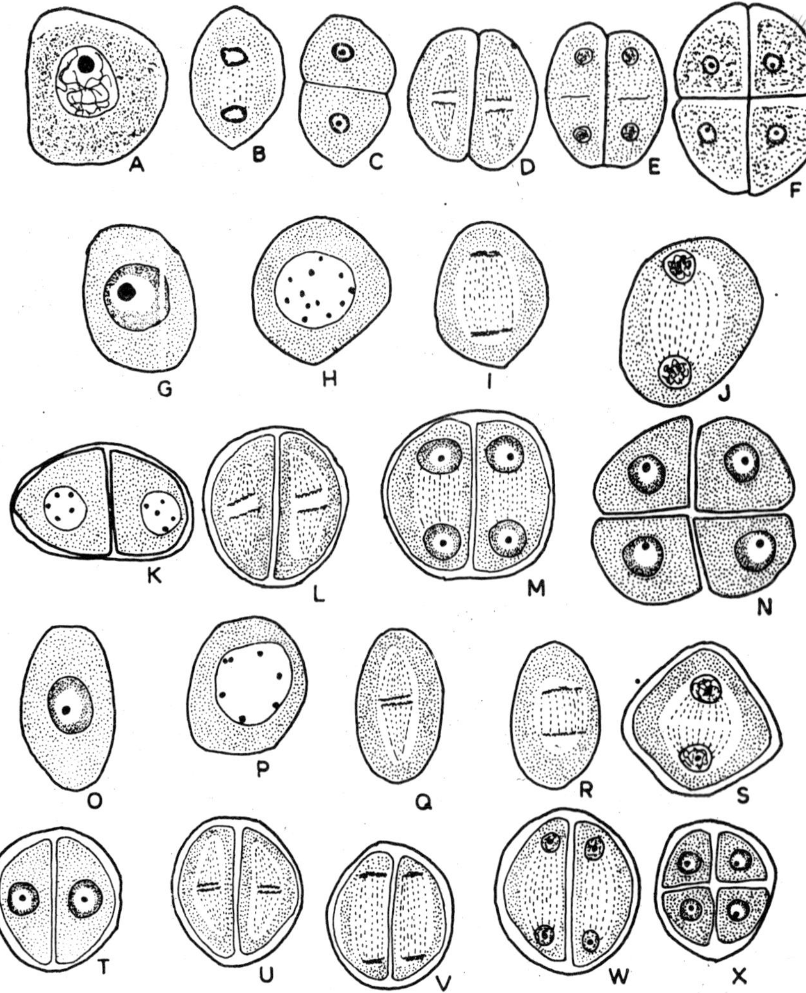

Fig. 4. Microsporogenesis. A-F. *Euchleana mexicana* (Koul, 1962). G-N. *Panicum notatum* Febulaus, 1992). O-X. *Brachiaria eruciformis* (Febulaus & Pullaiah, 1998).

(Jayalakshmi and Lakshmanan, 1982). In *Brachiaria reptans* tetrahedral microspore tetrads are occasionally observed (Febulaus and Pullaiah, 1991b).

Narayanaswami (1952, 1955c) while studying the microsporogenesis and male gametophyte in *Eleusine coracana* made some worthy observations. According to his observations the microspore tetrads are usually isobilateral or tetrahedral in arrangement. But linear, decussate and T-shaped tetrads were also observed in this species. In few cases less than four microspores are reported and this may be due to the failure of Division II in one of the dyad cell or one or sometimes two microspores of a tetrad become non-functional

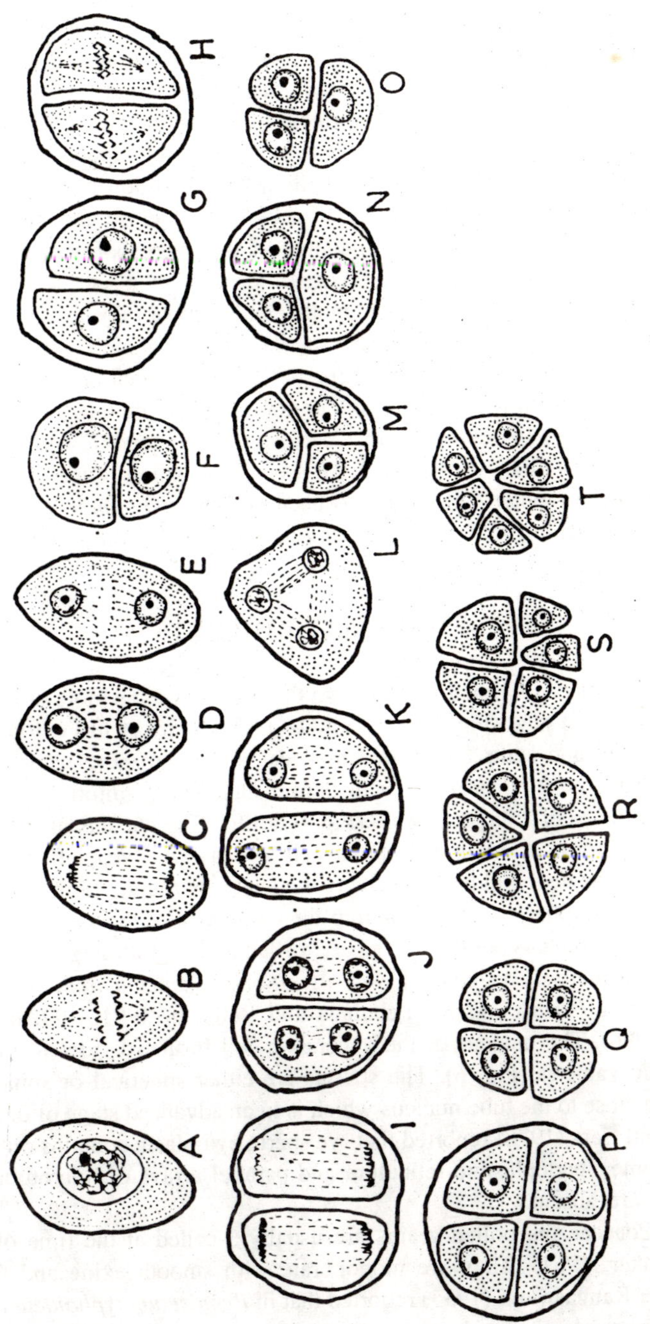

Fig. 5.A-L. Meiosis in pollen mother cells of apomictic species. M. Tetrahedral tetrad. N,O. Triads. P-Q. Pollen tetrads. R, S. Pentads, T. Hexads. A-T. *Cenchrus ciliaris* (Febulaus & Pullaiah, 1995).

and degenerate. Generally the microspores of a tetrad are of equal size and shape but in *Eleusine coracana* marked difference is noticed. One or two spores may be smaller or one may be considerably elongated than the others. This is especially noticed in T-shaped microspore tetrads. Usually the tetrads are free from one another when formed, but sometimes they show a tendency to remain united. Thus two tetrads may remain closely appressed to each other resulting in an association of eight cells. Such condition probably arises from the development of two pollen mother cells lying in close juxtaposition without separation. Variation in the arrangement of the microspores in such twins are also noticed (Narayanaswami, 1952).

Meiotic divisions in the pollen mother cells may or may not show synchrony. In *Eragrostis uniolodies* (Untawale *et al.*, 1969) complete synchronization is absent. In *Cenchrus ciliaris, Eragrostiella bifaria* meiotic divisions are not synchronous. Within the same anther lobes pollen mother cells at different meiotic stages were observed in *Cenchrus ciliaris,* and *Eragrostiella bifaria* (Febulaus and Pullaiah, 1992b, 1995). Asynchronization of meiotic divisions in the microspre mother cell have also been noticed in *Echinochloa frumentacea* (Narayanaswami, 1955) and *Spinifex littoreus* (Jayalakshmi and Lakshmanan, 1982).

MALE GAMETOPHYTE

The microspores after their release from the tetrads become more or less spherical and increase in size (Fig. 6 A–C,H,I,O–Q). They are richly cytoplasmic with a prominent and centrally located nucleus. The microspore nucleus moves to a peripheral position near the wall before undergoing mitotic division due to the appearance of a large vacuole (Fig. 6 D,J,R). The nucleus of the microspore divides to form a small generative cell and a large vegetative cell (Fig. 5 E,K,L,S). The wall laid between the two nuclei gets round off and the generative cell becomes spherical. It moves into the cytoplasm of the vegetative cell (Fig. 6 F,M,T). Vacuole is observed even at 2-celled pollen grain stage in *Cymbopogon* caesius, *Brachiaria reptans, Cenchrus ciliaris, Panicum notatum, Eragrostis viscosa* and *Perotis indica* (Febulaus and Pullaiah, 1991, 1992a,b, 1993a,b, 1995, 1996). The nucleus of the generative cell further divides to gives rise to two male gametes (Fig. 6). The sperms are either spherical or spindle-shaped. They lie close to the tube nucleus which is in an advanced stage of degeneration. Karas and Cass (1976) reported that rye sperm cytoplasm contains mitochondria, dictyosomes, endoplasmic reticulum and parallel arrays of microtubules, while plastids are absent.

In Poaceae the pollen grains are usually 3-celled at the time of shedding (Brewbaker, 1967). They are monoporate with smooth exine and thin intine. However Rangaswami (1935) reported that in *Pennisetum typhoideum* the pollen grains are shed at uninucleate condition. Narayanaswami (1953) while studying

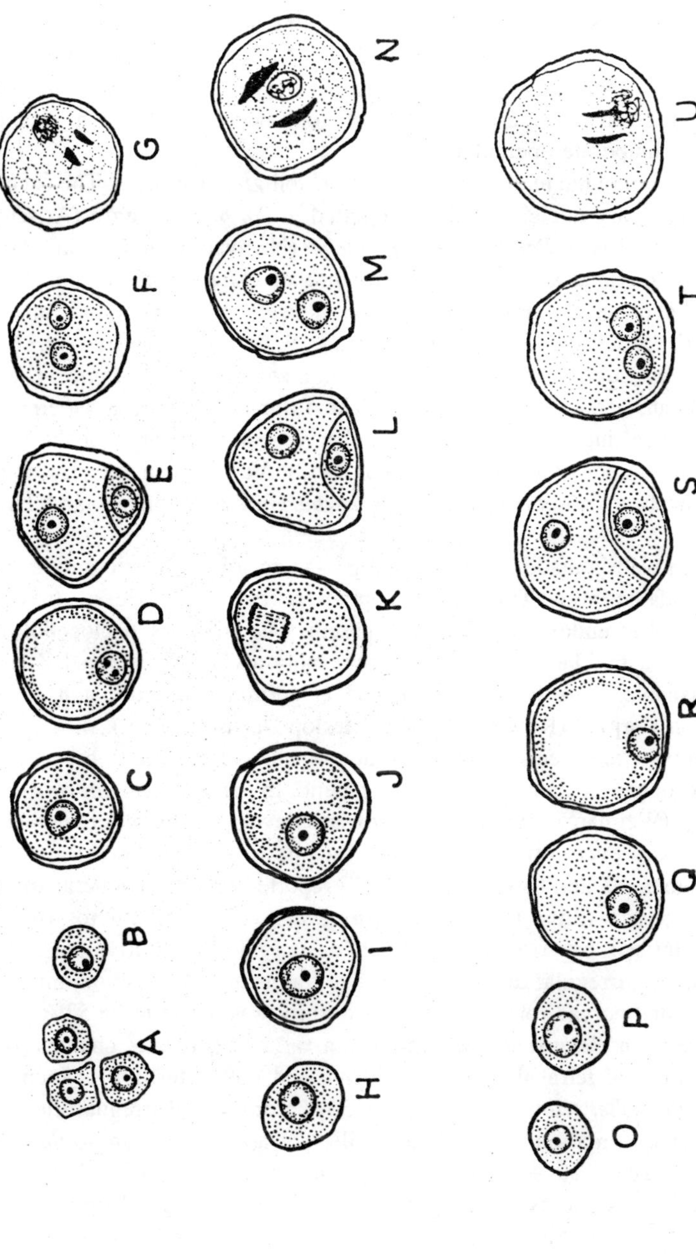

Fig. 6. A-U. Development of male gametophyte. A-G. *Brachiaria eruciformis* (Febulaus, 1991). H-N. *Chloris roxburghiana* (Febulaus and Pullaiah, 1991a). O-U. *Melanocenchris jacquemontii* (Febulaus & Pullaiah, 1997).

the structure and development of caryopsis in *Pennisetum typhoideum* stated that the pollen grains are 3-celled at the time of shedding rather than uninucleate as reported by Rangaswami (1935). In *Spinifex littoreus* Jayalakshmi and Lakshmanan (1982) reported that pollen grains are uninucleate at the time of shedding. Another exception was made by Chikkannaiah and Mahalingappa (1976c) in *Eleusine africana* and *E. compressa* where pollen grains are reported to shed at 2-celled stage. These observations need confirmation. The pollen grains are uniporate-operculate.

The nuclei in the mature pollen grain exhibit great variation. The occurrence of supernumerary nuclei have been reported in *Sorghum vulgare* (Artschwager and McGuire, 1949), *Pennisetum typhoideum* (Narayanaswami, 1953), *Panicum miliare* (Narayanaswami, 1955a), *Dicanthium pseudoischaemum, Panicum repens* (Febulaus and Pullaiah, 1992a), *Eleusine coracana* (Narayanaswami, 1952), *Eragrostis poaeoides* (Chandra, 1976), *Aristida hystrix, Chloris roxburghiana, Melanocenchris jacquemontii* and *Eragrostiella bifaria* (Febulaus and Pullaiah, 1991a, 1992b, 1993b). The formation of these supernumerary nuclei may be due to the division of either vegetative nucleus or the products of generative nucleus (Narayanaswami, 1952). Variation in the size of pollen grains has been observed in *Chloris roxburghiana* (Febulaus and Pullaiah, 1991a).

Pollen wall is made up of two layers—an acetolysis resistent exine and an acid degradable intine. The exine is composed of an outer ektexine (sexine and nexine 1). The endexine is virtually absent. The intine is pectocellulosic. In Gramineae, a middle layer, rich in pectic polysaccharides, termed Z-layer, is distinguishable. This layer is thickened at germinal aperture and is termed "Zwischen-Korper" (Rowley, 1964, J. Heslop-Harrison, 1979).

In apomictic species a number of abnormalities have been observed during microsporogenesis and pollen development. In *Hierochloe odorata* and *H. monticola* (Weimarck, 1967a,b) embryo sacs develop from pollen mother cells. In some cases even 4-nucleate embryo sacs have been recorded.

Shanthamma and Narayan (1976–77) reported that in apomictic species of *Cenchrus ciliaris* (2n = 45) and *Cenchrus glaucus* (2n = 45) microsporogenesis is characterised by irregularities such as univalents, multivalents, lagging chromosomes, irregular distribution, bridge fragment configuration, micronuclei formation and polyad formation. Pollen fertility was found to be 52% and 49% respectively. In *Pennisetum mezianum* (2n = 32), meiosis is characterised by the formation of tetravalents and bivalents and pollen fertility is found to be 92%. In *Rhynchelytrum villosum* (2n = 36) chromosomal irregularities lead to macronuclei formation and pollen fertility is about 60%. In *Anthoxanthum odoratum* (2n = 20 + 2) microsporogenesis is characterized by various abnormalities such as tetravalents, translocation rings, bridge fragment configuration and also 'B' chromosoms, pollen fertility was about 55% (Shanthamma and Narayan, 1976–77). In *Holcus mollis* Walter (1977) reported irregularities in microsporogenesis in polyploid forms (2n = 35). In addition to tetrads, triads, pentads, hexads and heptads are formed.

In *Eragrostis coarctata* (Chandra, 1976) though microsporogenesis proceeds in normal way, in a good number of plants meiotic irregularities have been noticed. These are lagging chromosomes, formation or micronuclei, polyad formation, degeneration of microspores, considerable variation in the size of the microspore in a microsporangium etc (Chandra, 1976). Such characters have also been reported for hybrids of *Zea mays* and *Tripsacum dactyloides* (Mangelsdorf and Reeves, 1939), hydrids between *Lolium perenne* and *Festuca pratensis* (Gymer and Wittingten, 1975a,b) and for certain other grasses (Church, 1929a,b).

In *Panicum notatum* (Febulaus, 1992), an apomictic species, pollen grains do not undergo any further development after 2-celled stage. The anther at maturity enclose only degenerated and sterile pollen grains without any cell contents. The pollen grains in this species are sterile.

In *Cenchrus ciliaris* triads, pentads and hexads (Fig. 4 N,O,R–T) have been observed. These might have resulted due to the meiotic abnormalities in pollen mother cells. The reproduction in this species is apomictic (Febulaus and Pullaiah, 1995).

One common cause of pollen abortion is the abnormal development and functioning of the anther tapetum, which either degenerates too early, as in wheat (Chauhan and Singh, 1965) or persists beyond its normal time as in *Sorghum* (Brooks *et al.*, 1966, Singh and Hadley, 1961), so that it fails to provide the developing sporogenous cells with such metabolites as deoxyribosides, several enzymes, sporopollen in globules and precursors of other wall substances.

Deshpande *et al.* (1981) said that the tapetum plays an important role in the development of pollen by providing nutrition to pollen mother cells and any deviation in the function of tapetal cells would cause premature death of pollen. In induced sterile-mutants of *Hordeum distichum* they found that the inner tangential walls of tapetal cells fail to get dissolved, may be due to non-secretion of certain enzymes required for the process, and as such fail to transfer the food material to the developing pollen mother cells, resulting in non-viable pollen grains i.e. male steriles.

Another cause of male sterility is the faulty digestion of callose which surrounds the meiocytes and young tetrads and isolates them from any harmful influence of the surrounding sporophytic tissue. Normally, its degradation starts with the onset of exine formation, but in several male-sterile lines the digestion is erratic. In cytoplasmic male-sterile lines of *Sorghum* (Damon, 1961) its dissolution is delayed. The precocious or delayed dissolution of callose results in abnormal pollen development and sterility.

Table 1. Structure and Development of Anther and Pollen

Species investigated and Author	Archesporium	Wall layers	Anther tapetum	Sporogenous tissue	Cyto-kinesis	Pollen tetrads	Pollen grains
Acrachne racemosa (Bhanwra, 1988)	—	Epidermis, endothecium, middlelayer and tapetum	Glandular, Cells 2-nucleate	—	Successive	Isobila-teral	3-celled
Acroceras munroanum (Shobha & Sindhe, 1992)	—	”	”	Moderately extensive	”	”	”
Agrositis pilosula (Muniyamma, 1976)	—	”	—	—	”	”	—
Alloteropsis cimicina (Sonpipare & Padhye, 1980)	1 or 2 rows	”	—	—	—	”	”
Andropogon pumilus (Kulkarni, 1982)	—	”	Glandular, cells uninucleate	—	”	”	3-celled
Apluda mutica (Bhanwra & Pathak, 1987)	—	”	”	36 sporogenous cells in vertical file	”	”	”
Aristida adscensionis (Bhanwra, 1988)	—	”	Glandular, cells 2-nucleate	—	”	”	
Aristida funiculata (Febulaus & Pullaiah, 1993b)	—	”	Glandular, cells 2-4-nucleate	2 rows of pollen mother cells	”	”	”
Aristida hystrix (Febulaus & Pullaiah, 1993b)	—	Epidermis, endothecium, middlelayer and tapetum	Glandular, Cells 2-nucleate	Single row of pollen mother cells	Successive type	Isobila-teral	3-celled

Aristida mutabilis (Febulaus & Pullaiah, 1993b)	—	”	Glandular, Cells 2-4 nucleate	Single row of pollen mother cells	Successive type	Isobilateral	3-celled
Aristida setacea (Febulaus & Pullaiah 1993b)	”	”	Glandular Cells 2-nucleate	2 rows of pollen mother cells	”	”	”
Arundinella mesophylla (Basappa & Muniyamma, 1981)	—	”	Glandular	—	”	”	—
A. nepalensis (Bhanwra, 1988)	—	”	Glandular Cells 2-nucleate	—	”	”	3-celled
A. purpurea (Basappa & Muniyamma, 1981)	Single celled	”	Glandular	—	”	”	—
Arundo donax (Bhanwra, 1988)	”	”	Glandular, cell 2-nucleate	—	”	”	3-celled
Avena fatua (Cannon, 1900)	”	—	—	—	—	—	—
Bambusa tulda (Bhanwra, 1988)	”	”	Glandular, Cell 32-nucleate	2 rows of pollen mother cells	Successive type	Isobilateral	3-celled
Bothriochloa odorata (Bhanwra, 1988)	”	Epidermis, endothecium, middle layer and tapetum	”	2 rows of pollen mother cells	Successive	Isobilateral	3-celled
Brachiaria distachya (Bhanwra, 1988)	—	Epidermis, endothecium, middle layer and tapetum	Glandular, cells uninucleate	2 rows of pollen mother cell	Successive	Isobilateral	3-celled

(*contd.*)

(Table 1 contd.)

Species investigated and Author	Archesporium	Wall layers	Anther tapetum	Sporogenous tissue	Cyto-kinesis	Pollen tetrads	Pollen grains
B. ramosa (Bhanwra *et al.*, 1985)	—	”	Glandular, cells 2-nucleate	”	”	”	”
B. reptans (Febulaus & Pullaiah 1991b)	Single-layered	”	”	Moderately extensive	”	”	”
Brachypodium sylvaticum (Bhanwra, 1988)	—	”	”	—	”	”	”
Bromus unioloides (Bhanwra, 1988)	—	”	”	—	”	”	”
Capillipedium huegelli (Bhanwra, 1988)	Single-layered	”	”	—	”	”	”
C. parviflorum (Bhanwra, 1988)	—	”	”	—	”	”	”
Cenchrus ciliaris (Gupta & Yashvir, 1971; Febulaus & Pullaiah, 1995)	—	Epidermis, endothecium, middle layer and tapetum	”	2-rows	Successive or simultaneous	Iso-bilateral or tetra-hedral or polyods	3-celled sterile
C. glaucus (Shanthamma, 1982)	—	”	Glandular	—	Successive	Iso-bilateral	3-celled
Chionachne koenigii (Satyamurty, 1984b)	—	—	Glandular cells 2-nucleate	One or two rows	”	”	Mono-colpate and 3-celled
Chloris gayana (Chikkannaiah & Mahalingappa, 1976)	One or two rows	”	”	Moderate mass	”	Isobi-lateral and decussate	3-celled

C. roxburghiana (Febulaus & Pullaiah, 1991a)	—	”	”	2 rows	”	”	3-celled
Chrysopogon fulvus (Sreenivasa Rao, 1997)	—	”	Glandular cells 2-nucleate	Moderate mass	Successive	Isobi-lateral	”
Cymbopogon caesius (Sreenivasa Rao, 1997)	—	Epidermis, endothecium, middle layer and tapetum	Glandular, cells 2-nucleate	Moderate mass	Successive or	Isobilateral, tetra-hedral	3-celled
Cymbopogon coloratus (Sreenivasa Rao, 1997)	—	”	Glandular, cells uninucleate	”	”	”	”
Cymbopogon flexuosus (Screenivasa Rao, 1997)	—	”	”	”	”	”	”
Cymbopogon martini (Choda *et al.*, 1982; Brown & Emery, 1958)	—	”	Glandular, cells 2-nucleate	4 to 5 rows of pollen mother cells	”	”	”
C. nardus (Choda *et al.*, 1982)	—	”	Glandular,	”	”	”	”
Cynodon dactylon (Mahalingappa, 1978)	1 row, Rarely a plate of cells in T.s.	”	Glandular	4 to 5 rows of pollen mother cells	Successive	Iso-bilateral and Decussate	3-celled
Dactyloctenium aegyptium (Chandra, 1963a; Venkateswarlu & Devi, 1964)	—	Epidermis, endothecium, middle layer and tapetum	Glandular	—	Successive	Iso-bilateral	3-celled
D. sindicum (Sharma *et al.* 1981)	1 row, Rarely a plate of cells in T.s.	Epidermis, endothecium, middle layer and tapetum	Glandular	4 to 5 rows of pollen mother cells	Successive	Isobi-lateral, and Decus-sate	3-celled

(contd.)

(Table 1 contd.)

Species investigated and Author	Archesporium	Wall layers	Anther tapetum	Sporogenous tissue	Cyto-kinesis	Pollen tetrads	Pollen grains
Dendrocalamus hamiltonii (Harigopal & Manasi Ram, 1981)	2 or 3 layered	,,	,,	—	,,	,,	,,
Desmostachya bipinnata (Bhanwra, 1988)	,,	,,	Glandular, Cells uninucleate	—	,,	,,	,,
Dicanthium armatum (Reddy & D' Cruz, 1969a)	—	,,	,,	—	,,	,,	,,
Digitaria adscendens (Kulkarni & Dnyansagar, 1981)	—	,,	,,	—	,,	,,	,,
D. bicornis (Venkateswarlu & Devi, 1964; Febulaus & Pullaiah, 1996)	—	Epidermis, Endothecium, middle layer and tapetum	Glandular, cell & uni-nucleate	—	Successive	Iso-bilateral	3-celled
D. ciliaris (Febulaus & Pullaiah, 1996)	—	,,	,,	2 rows of pollen mother cells	,,	,,	,,
Dinebra retroflexa (Satyamurty, 1985b)	—	,,	Glandular	—	,,	,,	,,
Echinochloa colonum (Bhanwra & Choda, 1986)	Single row	,,	Glandular, cells uninucleate	—	,,	Iso-bilateral, T-shaped, Decus-sate or intermediate types	,,

E. crusgalli (Bhanwra & Choda, 1986)	”	”	”	—	”	Isobilateral	”
E. frumentacea (Narayanaswami, 1955b)	—	”	Glandular, cells 2-nucleate	—	”	Isobilateral tetra-hedral, T-shaped and linear	”
E. stagnina (Muniyamma, 1978)	—	Epidermis, Endothe-cium, middle layer and tapetum	Glandular	—	Successive	Isobilateral, linear, T-shaped, decussate and inter-mediate types	3-celled
Eleusine compressa (Mahalingappa, 1977)	—	”	”	—	”	Iso-bilateral	2-celled
E. coracana (Narayanaswami, 1952)	—	”	Glandular, cells 2-nucleate	—	”	Iso-bilateral, T-shaped, Occasion-ally tetrahedral	3-celled, super-number-ary nuclei are present
E. indica (Chandra, 1963a)	—	Epidermis, Endothecium, middle layer and tapetum	Glandular, cells 2-nucleate	—	Successive	Iso-bilateral, T-shaped, Occasionally tetrahedral	—
Elytrophorus spicata (Satyamurty, 1985a)	Single row	”	”	—	”	Iso-bilateral, linear and T-shaped	3-celled
Eragrostiella bifaria (Venkateswarlu & Devi, 1964;	Single row	Epidermis, Endothe-cium, middle	Glandular, cells 2-nucleate	2 rows	Successive	Isobilateral	3-celled

(contd.)

(Table 1 contd.)

Species investigated and Author	Archesporium	Wall layers	Anther tapetum	Sporogenous tissue	Cyto-kinesis	Pollen tetrads	Pollen grains
Febulaus & Pullaiah, 1992b)		layer and tapetum					
Eragrostis ciliaris (Bhanwra, 1986a)	—	”	”	—	”	”	”
E. ciliensis (Kulkarni & Dnyansagar, 1984)	—	Epidermis, Endothecium, middle layer and tapetum	”	—	Successive	Iso-bilateral	3-celled
E. coarctata (Venkateswarlu & Devi, 1964)	—	”	Glandular, cells 1-nucleate	2-3 rows	”	”	”
E. diarrhena (Venkateswarlu & Devi, 1964)	—	”	Glandular, cells 2-nucleate	—	”	”	”
E. minor (Bhanwra, 1988a)	—	”	”	—	”	”	”
Eragrostis namquensis (Ghaisas & Patil, 1990)	—	”	”	—	”	”	”
E. neesii (Venkateswarlu &) Devi, 1964)	—	”	Glandular, cells 1-nucleate	2-3 rows	”	”	”
E. nigra (Bhanwra, 1986)	—	”	”	—	”	”	”
E. plumosa (Venkateswarlu & Devi, 1964)	—	”	”	—	”	”	”
E. pilosa	—	”	Glandular,	—	”	”	”

(Chandra, 1976)			cells 2-nucleate				
E. poaeoides (Bhanwra, 1987a; Chandra, 1976)	—	”	”	—	”	”	”
Eragrostis tenella (Venkateswarlu & Devi, 1984; Kalantri, 1990	—	”	”	—	”	”	”
E. tremula (Bhanwra, 1986a)	—	”	”	—	”	”	”
E. unioloides (Untawale *et al.* 1968; Chandra, 1976)	Single row	”	”	Moderate mess	”	”	”
E. viscosa (Venkateswarlu & Devi, 1964, Febulaus & Pullaiah, 1990)	”	”	”	Single row	”	”	”
Eremopogon foveolatus (Seshavatharam & Bhaskara Rao, 1988)	—	Epidermis, Endothe-cium, middle layer and tapetum	Glandular	—	Successive	Iso-bilateral	3-celled
Euchlaena mexicana (Koul & Sahi, 1968)	—	”	Glandular cells 2-nucleate	2 rows	”	”	”
Glyceria tonglensis (Bhanwra, 1988)	—	”	Glandular, cells 2-nucleate	—	”	”	”
Helictotrichon virescens (Bhanwra, 1988)	—	Epidermis, Endothe-cium, middle layer and tapetum	Glandular cells 2-nucleate	—	Successive	Iso-bilateral	3-celled

(contd.)

(Table 1 contd.)

Species investigated and Author	Archesporium	Wall layers	Anther tapetum	Sporogenous tissue	Cyto-kinesis	Pollen tetrads	Pollen grains
Imperata cylindrica (Bhanwra, 1988)	—	”	”	—	”	”	”
Ischaemum rugosum (Sreenivasa Rao, 1997)	—	”	Glandular cells 1-nucleate	Moderate mass	Successive	Iso-bilateral	3-celled
Iseilema anthephoroides (Venkateswarlu & Devi, 1974)	—	”	Glandular cells 2-nucleate	—	”	”	”
Iseilema prostratum (Bhanwra, 1981)	—	”	”	—	”	”	”
Leersia hexandra (Venkateswarlu & Devi, 1964)	—	”	”	—	”	”	”
Leptechloa chinensis (Bhanwra *et al.*, 1981)	—	”	”	1 or 2 rows	”	”	”
L. neesii (Venkateswarlu & Devi, 1964)	—	”	”	”	”	”	”
L. panicea (Venkateswarlu & Devi, 1964; Bhanwra, 1988)	”	”	”	—	”	”	”
Limnopoa meeboldii (Shobha & Sindhe, 1996)	Single row	”	”	Moderately extensive	”	”	”
Lolium perenne (Bhanwra, 1988)	—	Epidermis, Endothecium, middle layer and tapetum	Glandular cells 2-nucleate	1 or 2 rows	Successive	Iso-bilateral	3-celled

Melanocenchris jacquemontii (Febulaus & Pullaiah, 1997)	Single row	”	”	2 rows	”	”	”
Microstegium ciliiatum (Bhanwra, 1988)	—	”	”	2 rows	”	”	”
Oplismenus burmanii (Bhanwra, 1988)	—	”	”	—	”	”	”
O. compositus (Bhanwra, 1988)	—	”	”	—	”	”	”
Oropetium thomaeum (Diwanji & Diwanji, 1981)	Single layer	”	Glandular, cells uninucleate	—	”	”	”
O. villosulum (Diwanji & Diwanji, 1981)	”	”	”	—	”	”	”
Orzya alta (Sreenivasa Rao, 1997)	Single layer	Epidermis, Endothe-cium, middle layer and tapetum					
Oryza brachyantha (Sreenivasa Rao, 1997)	Single row	Epidermis, Endothe-cium, middle layer and tapetum	Glandular cells 1-nucleate	Moderate mass	Successive	Isobi-lateral, Decus-sate	3-celled
Oryza longistamina (Sreenivasa Rao, 1997)	”	”	”	”	”	Isobi-lateral, T-shaped	”
Oryza malampuzhaensis (Sreenivasa Rao, 1997)	”	”	”	2 rows	”	Isobi-lateral, tetrahedral	”

(contd.)

(Table 1 contd.)

Species investigated and Author	Archesporium	Wall layers	Anther tapetum	Sporogenous tissue	Cyto-kinesis	Pollen tetrads	Pollen grains
Oryza minuta (Sreenivasa Rao, 1997)	”	”	”	”	”	”	”
Oryza officinalis (Sreenivasa Rao, 1997)	”	”	”	”	”	”	”
Oryza rhizomatis (Sreenivasa Rao, 1997)	”	”	”	”	”	”	”
Oryza ridleyi (Sreenivasa Rao, 1997)	”	”	”	”	”	”	”
Oryza rufipogon (Sreenivasa Rao, 1997)	”	”	”	”	”	”	”
O. sativa (Bhanwra, 1988; Raju, 1980; Sapre, 1964, 76)	Two rows	Epidermis, Endothe-cium, middle layer and tapetum	Glandular cells 2-nucleate	3-5 rows	”	”	”
Oryzopsis miliacea (Maze *et al.*, 1971)	Single row	”	Glandular	—	”	”	”
Panicum miliaceum (Narayanaswamy, 1955a)	—	Epidermis, Endothe-cium, middle layer and tapetum	Glandular cells 2-nucleate	—	Successive	Iso-bilateral	3 celled

P. miliare (Narayanaswami, 1955a)	—	”	Glandular, cells uninucleate	—	”	Iso-bilateral, or tetra-	3-celled, show super-number-ary nuclei
P. notatum (Febulaus, 1992)	—	”	Glandular, cells 2-nucleate	2 rows	”	Iso-bilateral	2-celled sterile pollen grains
Panicum repens (Febulaus & Pullaiah, 1992a)	Single rows	”	”	3 layers of pollen mother cells	”	”	”
Paspalidium flavidum (Bhanwra, 1988)	—	”	Glandular, cells 2-nucleate	—	”	”	3-celled
Paspalum conjugatum (Schnarf, 1929)	—	Epidermis, Endothe-cium, middle layer and tapetum	Glandular,	—	Successive	Iso-bilateral	3-celled
P. distichum (Bhanwra, 1988)	—	”	Glandular, cells 2-nucleate	—	”	”	”
P. mandiocanum (Schnarf, 1929)	—	Epidermis, Endothe-cium, middle layer and tapetum	Glandular	—	Successive	Iso-bilateral	3-celled

(contd.)

(Table 1 contd.)

Species investigated and Author	Archesporium	Wall layers	Anther tapetum	Sporogenous tissue	Cyto-kinesis	Pollen tetrads	Pollen grains
Pennisetum orientale (Bhanwra, 1988)	—	”	Glandular, cells 2-nucleate	—	”	”	”
Pennisetum pedicellatum (Shobha, 1988)		”	Glandular cells 2-nucleate	—	”	”	”
Pennisetum typhoideum (Narayanaswami, 1953; Raju, 1980)	2-3 cells	”	Glandular cells 1-nucleate	4-7 rows	”	”	3-celled some-times 4-5-celled
Perotis hordeiformis (Venkateswarlu & Devi, 1964)	—	Epidermis, Endothe-cium, middle layer and tapetum	Glandular, cells 2-nucleate	—	”	”	”
Phalaris minor (Bhanwra, 1988)	—	”	”	—	”	”	”
Phragmites communis (Satyamurty & Seshavatharam, 1984)	—	”	”	—	”	”	Sterile
P. karka (Bhanwra *et al.*, 1988)	—	”	”	—	”	”	—
Poa annua (Bhanwra *et al.*, 1988)	—		Glandular, cells 1-nucleate	—	Successive	Iso-bilateral	3-celled
P. secunda (Kellogg, 1987)	—	”	Glandular,	—	”	”	”
Polypogon fugax (Bhanwra *et al.*, 1981)	—	”	Glandular, cells 2-nucleate	1 or 2 rows	”	”	”

P. monspeliensis (Bhanwra *et al.*, 1981)	—	”	”	—	”	”	
Rottboellia exaltata (Sreenivasa Rao, 1997)	—	”	Glandular cells 2-nucleate	Moderate mass	Successive	Isobi-lateral	3-celled
Rostraria phleoides (Bhanwra, 1988)	—	”	”	—	”	”	”
Saccharum benghalense (Bhanwra, 1988)	—	”	”	—	”	Iso-bilateral and decussate	”
Setaria glauca (Bhanwra, 1988)	—	”	Glandular, cells 1-nucleate	—	”	”	”
S. intermedia (Bhanwra *et al.*, 1980)	—	”	”	—	”	”	”
Setaria interrupta (Sreenivasa Rao, 1997)	1 or 2 rows	Epidermis, Endothe-cium, middle layer and tapetum	Glandular cells 2-nucleate	Moderate mass	Successive	Isobilateral	3-
						celled	
S. italica (Khosla, 1948;) Narayanaswami, 1956	—	”	Glandular, cells 1-nucleate	2 rows	”	Isobilateral or tetrahedral	
Setaria pumila (Sreenivasa Rao, 1997)	1 or 2 rows	”	”	Moderate mass	”	Isobi-lateral	”
S. verticillata (Bhanwra *et al.* 1980)	—	”	Glandular, cells occasionally 2-nucleate	—	”	Iso-bilateral, decussate, linear, T-shaped or isolateral	3- celled

(contd.)

(Table 1 contd.)

Species investigated and Author	Archesporium	Wall layers	Anther tapetum	Sporogenous tissue	Cyto-kinesis	Pollen tetrads	Pollen grains
Sorghum arundinaceum (Venkateswarlu & Devi, 1964)	—	”	”	—	”	Iso-bilateral	”
Spinifex littoreus (Lakshmanan & Jayalakshmi, 1980; Jayalakshmi & Lakshmanan, 1982)	Single row	”	Glandular, cells uninucleate	Moderate mass	”	Iso-bilateral, linear, tetrahedral and decussate	”
Sporobolus coromandelianus (Seshavatharam & Satyamurty, 1982a)	—	”	Glandular cells 2-nucleate	—	”	Isobilateral	”
S. diander (Bhanwra *et al.*, 1981)	—	Epidermis, Endothe-cium, middle layer and tapetum	Glandular, cells 2-nucleate	1 or 2 rows	Successive	Isobilateral	3-celled
S. fertilis (Bhanwra, 1988)	—	”	Glandular cells 2-nucleate	—	”	”	”
Sporobolus indicus (Sreenivasa Rao, 1997)	—	”	Glandular cells 2-nucleate	Moderate mass	”	Isobilateral	”
Sporobolus spicatus (Srinivasa Rao, 1997)	—	”	”	”	”	Isobilateral, tetrahedral	3-celled
S. tremulus (Seshavatharam & Bhaskara Rao, 1983)	—	”		—	”	”	”

Stipa tortilis (Maze *et al.*, 1971)	Single row	”	Glandular	—		”	”
Thyrsostachys oliveri (Bhanwra *et al.*, 1988)	—	”	Glandular, cells 1-nucleate	—	”	”	”
Trachys muricata (Venkateswarlu & Devi, 1964)	—	”	Glandular, cells	—	”	”	”
Tripogon filiformis (Bhanwra *et al.*, 1981)	—	”	Glandular, cells 1-nucleate	1 or 2 rows	”	”	”
Triticum aestivum (Bhatnagar & Chandra, 1976; Raju, 1980)	Single row	”	Glandular, cells-2 nucleate	Many rows	”	”	”
T. dicoccum (Raju, 1980)	”	”	”	”	”	”	”
T. durum (Raju and Deshpande, 1981; Raju, 1980)	”	”	”	—	”	”	”
Urochloa panicoides (Basavaiah and Murthy, 1989)	—	Epidermis, Endothe-cium, middle layer and tapetum	Glandular	—	Successive	Iso-bilateral, rarely T-shaped	3-celled
Vetiveria zizanioides (Bhanwra, 1988)	—	”	Glandular	—	”	”	”
Zea mays (Raju, 1980)	2-3 rows	”	Glandular cells 2-nucleate	3-6 rows	”	—	”
Zingeria trichopoda (Shehata, 1995)	—	”	Glandular cells 2-nucleate	Six rows	”	Isobilateral	”

CHAPTER 3

MEGASPORANGIUM AND MEGASPOROGENESIS

The ovary in the family Poaceae is unilocular with a single basal ovule attached to its posterior wall. In *Spinifex littoreus* sometimes two ovules of different sizes develop, the second ovule is embedded in the ovarian tissue at the base of the style (Jayalakshmi and Lakshmanan, 1982).

The ovule arises as a papillate out growth at the base of the ovary (Fig. 7 A). During further growth the ovule undergoes a curvature towards the base of the ovary and finally becomes hemianatropous (Fig. 7 B–H,J,K). The ovule in the family Poaçeae has been described variously in different members. In majority of the members the ovule is hemianantropous or campylotropous. In addition to these anatropous, anacampylotropous, amphitropous and orthotropous ovules have also been reported.

Campylotropous ovules have been reported in *Pennisetum typhoideum* (Narayanaswami, 1953), *Eleusine coracana* (Narayanaswami, 1955c), *Sporobolus coromandelianus* (Satyamurthy and Seshavatharam, 1984), *Elytrophorus spicata* (Satyamurthy, 1985a), *Eragrostis unioloides* (Untawale *et al.*, 1969), *E. ciliaris, E. nigra, E. poaeoides, E. tenella, E. tremula* (Bhanwra, 1986a) and *Demostachya bipinnata* (Bhanwra, 1986b).

In *Paspalum scrobiculatum* (Narayanaswami, 1954), *Echinochloa frumentacea* (Narayanaswami, 1955b), *Setaria itatica* (Narayanaswami, 1965), *Triticum aestivum* (Bhatnagar and Chandra, 1975) and *Spinifex littoreus* (Jayalakshmi and Lakshmanan, 1982) anatropous ovules have been reported.

Calderson and Soderstrom (1973) described the ovule in *Maclurolyra* as anacampylotropous. Amphitropous ovules have been reported in *Euchlaena mexicans, Zea mays* (Cooper, 1937), *Eustachys petraea* and *E. glauca* (Aulbach-Smith and Herr 1984). Bhanwra *et al.* (1985) reported orthotropous ovules in *Poa annua.*

The ovule in the family Poaceae has been described as bitegmic (Figs. 7 E–H,J,K, 8 A–J, 9 H,J, 10 K, 11F) (Narayanaswami, 1952, 1953, 1954, 1955, 1956; Chandra, 1963; Venkateswarlu and Devi 1964; Maze and Bohm, 1974; Maze *et al.*, 1972; Bhanwra *et al.*, 1985; Bhanwra, 1988; Febulaus and Pullaiah, 1991a,b, 1992a,b, 1993a,b, 1995, 1996). The integuments initiate

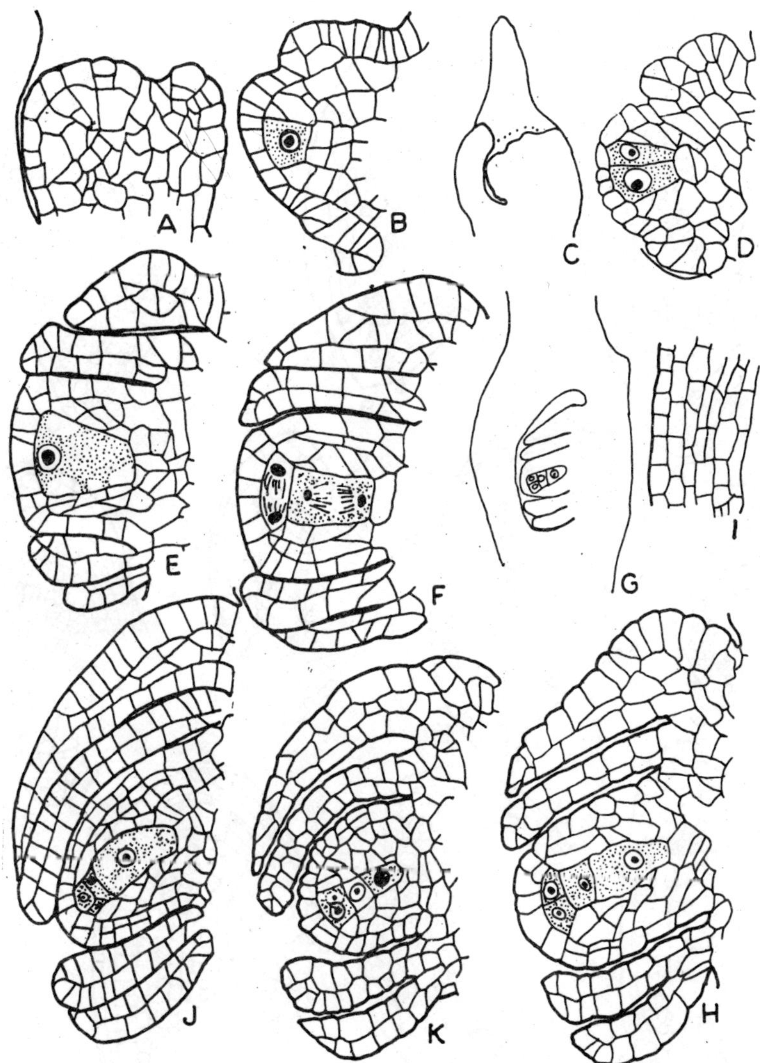

Fig. 7. A-H. Stages in the development of ovule in *Stipa elmeri* (Maze & Bohm, 1973).

through periclinal divisions in the protoderm. Both the integuments initiate at the same time on the lower side of the ovule. The inner integument is longer than the outer integument and grows faster than the outer integument. At the time of early development the inner integument is two to three layered in thickeness. By megaspore stage an internal layer of cells appears at the micropylar end. This thickened portion of the inner integument delimits the micropyle (Maze *et al.*, 1972). The outer integument is two cells in thickness.

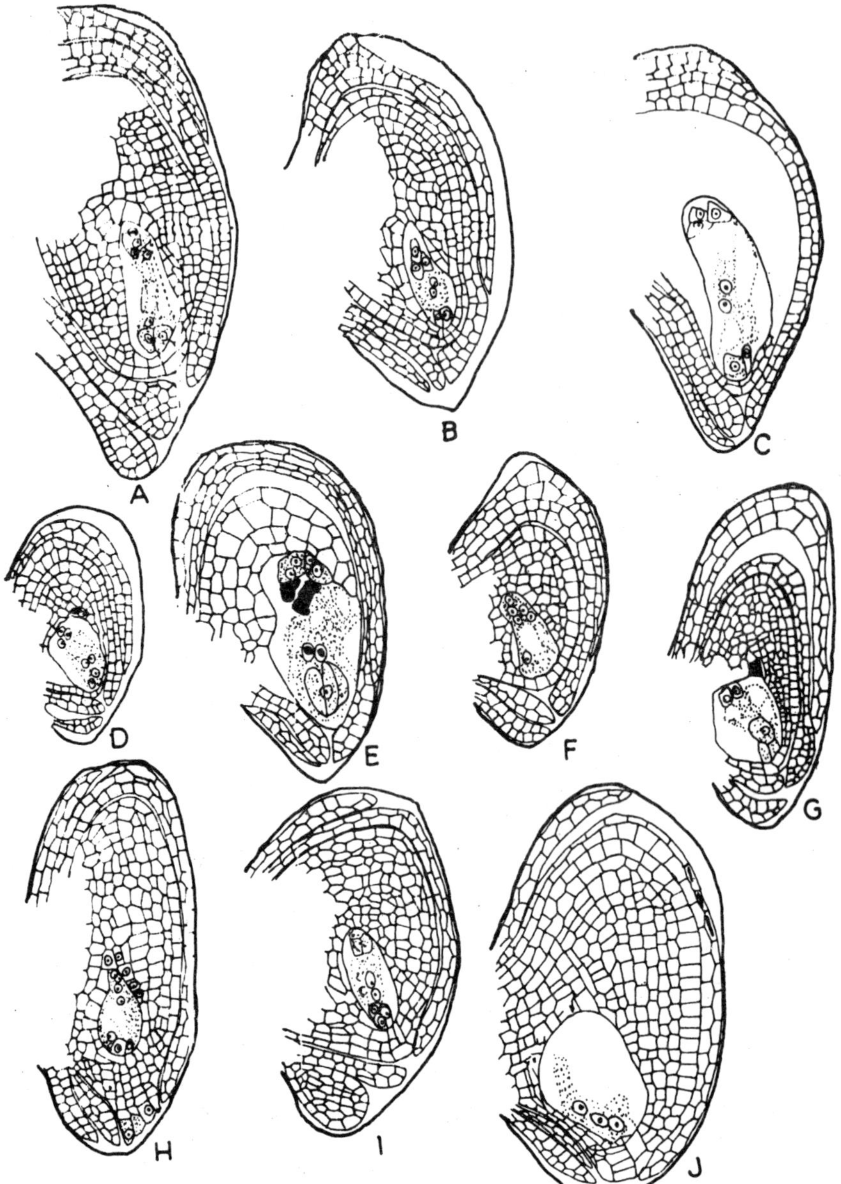

Fig. 8. V.s. of mature ovules. A. *Aristida adscensionis.* B. *Brachypodium sylvaticum.* C. *Glyceria tonglensis.*D. *Leptochloa panicea.* E. *Chloris barbata.* F. *Tragus roxburghii.* G. *Perotis hordeiformis.* H. *Arundinella nepalensis.* I. *Saccharum bengalense.* J. *Capillipedium huegellii* (Bhanwra, 1980).

The extent of growth of outer integument differs markedly in the two subfamilies of Poaceae. In Pooideae both the integuments are well developed.

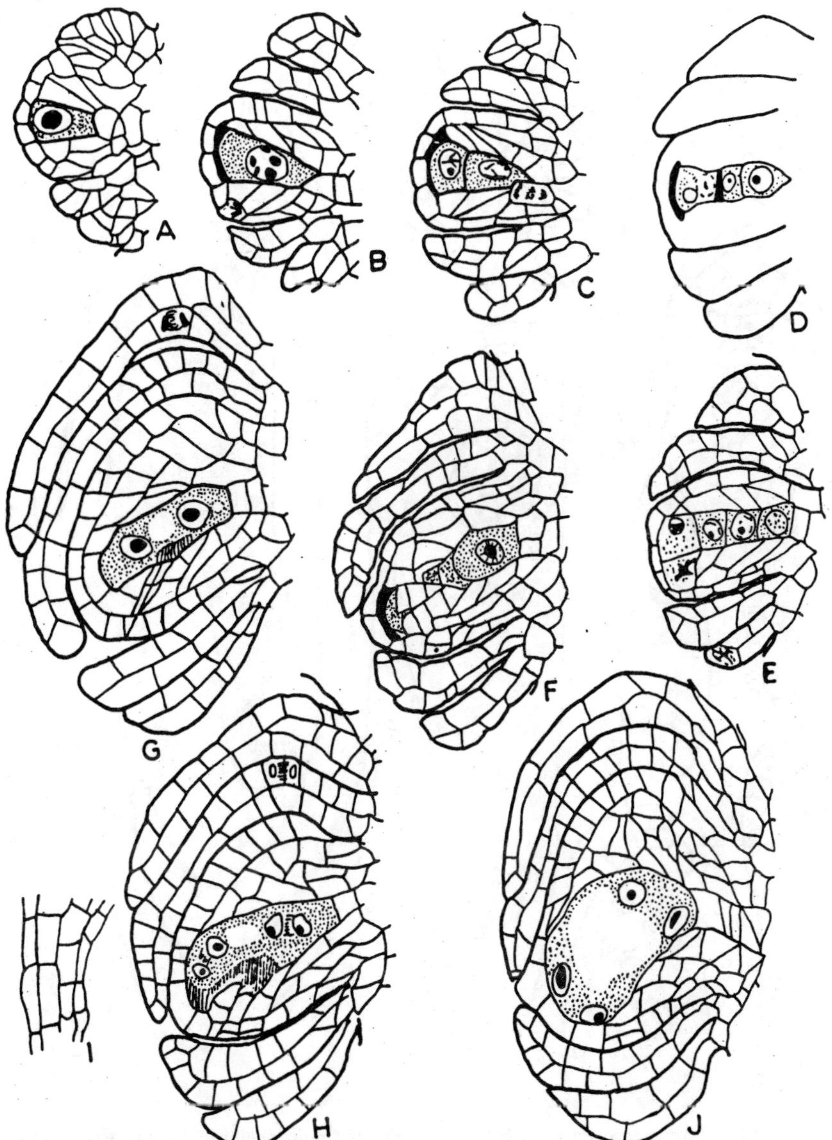

Fig. 9. A-J. Stages in the development of ovule and embryo sac in *Agrostis interrupta* (Maze & Bohm, 1974).

The outer integument invests the ovule for the greater part except the micropylar region. In Panicoideae, on the other hand the outer integument extends only up to about half the length of the ovule (Chandra, 1963; Verboom *et al.*, 1994).

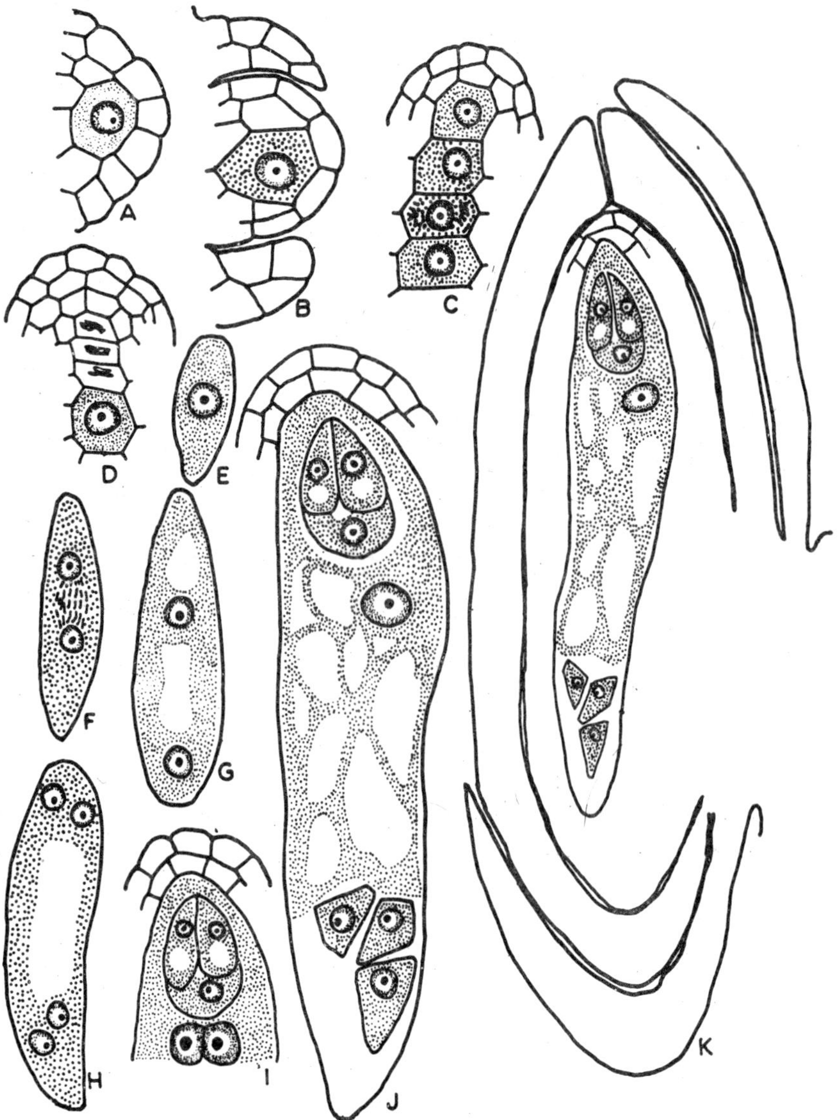

Fig. 10. A–K. Megasporogenesis and female gametophyte development in *Digitaria bicornis* (Febulaus & Pullaiah, 1996).

Venkateswarlu and Devi (1964) reported that the outer integument in some grasses covers only less than 2/3rd of the ovule. According to Bhanwra (1988) the outer integument is poorly developed and encloses less than half the length of the ovule in the Panicoid members.

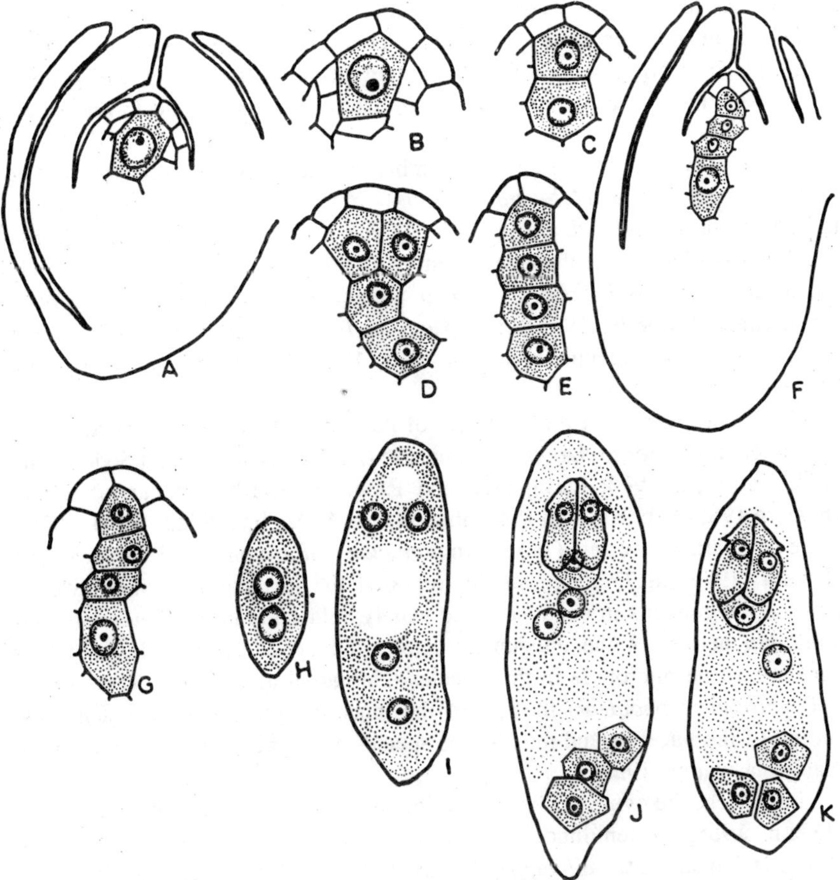

Fig. 11. Megasporogenesis and female gametophyto development in *Eragrostiella bifaria* (Febulaus & Pullaiah, 1992b).

Exception to the bitegmic condition is reported in *Spinifex littoreus* by Jayalakshmi and Lakshmanan (1982) where the ovule is described as unitegmic. They regarded the outer integument as obturator. The envelop described as obturator in *Spinifex littoreus* by Jayalakshmi and Lakshmanan (1982) has been interpreted as the outer integument by Narayanaswami (1952, 1953, 1955a,b,c); and Maze and Bohm, (1973, 1974). According to Jayalakshmi and Lakshmanan (1982) the outer envelop neither develops immediately after the initiation of the inner integument nor form the nucellar region. More over when the ovule is making a shift on its position during its ontogeny the outer envelop is not involved in it. This envelop stops as a hood over the micropyle and extends as an out growth at the chalazal region and is not in any way related to the position of the micropyle. Therefore the paucity of time in its origin, place of origin and its incomplete development favoured to designate it as obturator rather than as an outer integument (Jayalakshmi and Lakshmanan, 1982).

There is variation in size and number of integuments in bamboos. The ovules are ategmic in *Melocanna bambusoides* (Stapf, 1904; Petrova, 1965) and unitegmic in *Melocalamus* and *Ochlandra*. They are bitegmic in *Bambusa* and *Dendrocalamus* but the integument fail to cover the nucellus completely and no micropyle is formed. The outer integument in *Phyllostachys nigra* is rudimentary (Gioelli, 1935). In *Arundinaria* and *Maclurolyra* the inner integument forms the micropyle (Yoshida, 1963; Calderon and Soderstrom, 1973).

The micropyle in all the members of the family Poaceae is formed by the inner integument alone (Narayanaswami, 1954; 1955c; Calderon and Soderstrom, 1973; Philipson 1978; Deshpande and Raju, 1979; Bhanwra *et al.*, 1982; Bhanwra 1988; Febulaus and Pullaiah, 1991a,b, 1992a,b, 1993a,b, 1995, 1996). However Narayanaswami (1955a) in *Panicum miliaceum* and Bhatnagar and Chandra (1975) in *Triticum vulgare* reported the formation of micropyle by both the integuments in contrast to the widely held opinion. In *Dendrocalamus hamiltonii* neither of the integuments cover the tip of the nucellus and thus there is no true micropyle in this species (Harigopal and Manasi Ram, 1981). In *Pennisetum typhoideum* the rim of the inner integument never reaches the summit of the nucellus and thus in this species also typical micropyle is absent (Narayanaswamy, 1953).

In Poaceae the ovule has been described as crassinucellate (Narayanaswami, 1955a,b, 1956) or tenuinucellate (Narayanaswami, 1955c; Chandra, 1976; Chikkannaiah and Mahalingappa, 1976a,b). Davis (1966) has introduced the term pseudocrassinucellate to account for the periclinal divisions in the nucellar epidermis.

In the subfamily Panicoideae the ovule has been reported as Pseudo-crassinucellate (Chandra, 1963b; Bhanwra, 1988). In this sub-family the nucellar epidermis divides periclinally to form well developed parietal tissue in the micropylar region. Presence of two to seven layers in the vicinity of the micropylar region is of regular occurrence in the members of the Panicoideae (Bhanwra, 1988). In Pooideae, the nucellar epidermis below the micropyle usually does not divide periclinally. Consequently the ovules in Panicoideae are described as pseudocrassinucellate and in Pooideae as tenuinucellate (Chandra, 1963; Maze *et al.*, 1970; 1972; Bhanwra *et al.*, 1981; Bhanwra, 1988; Satyamurthy, 1985a,b). However variation has been met within the subfamily Pooideae. Periclinal division in the nucellar protoderm at the micropylar region are reported in the *Stipeae* (Maze *et al.*, 1970; 1972; Maze and Bohm, 1973; Mehlenbacher, 1970). In *Aristida adscensionis* (Bhanwra *et al.*, 1982), *Aristida hystrix* and *A.*

mutabilis (Pooideae) (Febulaus and Pullaiah, 1993b), the nucellar epidermis undergoes one or two periclinal divisions in the vicinity of micropyle to form one or two parietal layers.

In *Spinifex littoreus* a member of the tribe Paniceae of Panicoideae the parietal tissue is of dual origin. Both the nucellar epidermis and the parietal cell divide periclinally to add the number of layers in the micropylar region. About 13 to 15 such parietal layers have been observed in this species (Jayalakshmi and Lakshmanan, 1982).

Nucellus is generally absorbed by the developing endosperm. A definite perisperm layer of tangentially elongated and persistent nucellar cells has however been recorded in *Oryza sativa* (Santos, 1933) and *Euchlaena mexicana* (Cooper, 1937).

The nucellar cells below the megaspore mother cell in *Triticum* differentiate as an axial row of tubular cells. They are in close contact with the double connective bundle and probably have a role as conductive tissue.

Seshavatharam and Satyamurthy (1982b) in their study on 24 species of grasses belonging to 12 different tribes of Poaceae reported the occurrence of Hypostase. Just at the level of origin of the integuments and below the embryo sac occurs a well defined group of nucellar cells forming the hypostase. The cells of the hypostase persist in the mature caryopsis as a cushion in between the placental region of the ovule and endosperm. The parietal layer of the endosperm adjoining the hypostase becomes the transfer aleurone layer after undergoing elongation and specialization with striations. The occurrence of placental zone and this specialized aleurone layer on either side is a positive indication for the nutritive role played by the hypostase.

An Epistase is conspicuous in *Themeda australis* (Woodland, 1964).

MEGASPOROGENESIS

The female archesporium is hypodermal in all the members of the family Poaceae. However in *Dicanthium armatum* (Reddy and D'Cruz, 1969c) and *D. pseudoischaemum* (Febulaus and Pullaiah, 1991) archesporium develops in the central region of the nucellus. In *Bouteloua curtipendula* complex, Mohamed and Gould (1966) reported that the megasporocyte is differentiated deep in the nucellus.

The female archesporium is usually unicellular in several members of the family (Chandra, 1963a; Narayanaswami, 1953, 1954, 1956). Variation has been met within some species. In *Eragrostis ciliensis* (Swamy, 1944; Kulkarni and Dnyansagar, 1984) the female archesporium is 2-celled. Bhanwra *et al.* (1982) reported frequent occurrence of 2-celled female archesporium in *Vetiveria zizanioides.* In *Thyrsostachys oliveri* (Bhanwra, 1989), *Stipa elmeri* (Fig. 7D) (Maze and Bohm, 1973; Maze and Lin, 1975), *Themeda cymbaria* (Muniyamma, 1973), *Saccharum benghalense* (Bhanwra *et al.*, 1982), *Poa compressa*

(Anderson, 1927), *Oryzopsis miliacea* (Maze *et al.*, 1971), the female archesporium is usually single-celled but occasionally it is 2-celled. In *Chionachne koenigii* (Satyamurthy, 1984b), *Hierochloe australis, H. alpina* (Weimarck, 1967a; 1970) it is 1–3-celled. In *H. odorata* and *H. monticola* (Weimarck 1967a, 1970) it is 1–5-celled. Multicellular archesporium has been reported in *Eleusine compressa* (Mahalingappa, 1977), *Festuca microstachys* (Maze and Bohm, 1977) and in *Oryzopsis hendersonii* (Mehlenbacher, 1970).

The female archesporium is differentiated even before the initiation of integuments in *Dactyloctenium aegyptium* (Chandra, 1963a), *Stipa elmeri* (Maze and Bohm, 1973), *Triticum aestivum* (Aziz, 1972) and *Elytrophorus spicata* (Satyamurthy, 1985a).

The archesporial cell enlarges considerably and directly functions as megaspore mother cell. However Jayalakshmi and Lakshmanan (1982) reported that the archesporial cell undergoes periclinal division resulting in a parietal cell and sporogenous cell.

The megaspore mother cell enlarges considerably and undergoes meiotic division. In all the sexually reproducing species the meiotic division is normal. The heterotyphic division is always transverse and gives rise to two dyad cells. The homotyphic division is also transverse and results in a linear tetrad of megaspores (Figs. 9E, 10C,D, 11E). Sometimes the micropylar dyad cell divides in a plane at right angles to that of the chalazal one resulting in a 'T'-shaped megaspore tetrad (Figs. 7H, 11D).. Thus the megaspore tetrads in Gramineae are either linear or T-shaped (Narayanaswami, 1952; 1953; 1954; 1955a,b,c; 1956; Chandra, 1963a; Venkateswarlu and Devi, 1964; Maze *et al.,* 1970 1972; Maze and Bohm, 1973; Muniyamma, 1978; Choda *et al.*, 1982; Bhanwra 1985; 1988 etc.) In *Dicanthium armatum* Reddy and D'Cruz (1969c) reported tetrahedral megaspore tetrads in addition to the normal T-shaped or linear tetrads.

The size and shape of the dyad cells varies slightly. In *Eustachys petraea* the micropylar dyad member is much smaller than the chalazal dyad. Meiosis II occurs non-synchronously in the dyad members. The micropylar division is lagging behind that of the chalazal dyad cell. Occasionally, cytokinesis does not occur after meiosis II in the upper dyad member thus producing a modified T-shaped tetrad (Aulbach-Smith and Herr, 1984). In *Saccharum benghalense* (Bhanwra, 1988) the division in the upper dyad was not accompanied by wall formation.

Maze and Bohm (1974) reported that the first meiotic division results in a large chalazal cell and the second meiotic division gives rise to a linear tetrad of megaspores or a modified T-shaped tetrad without walls between the two micropylar megaspores. In *Brachiaria eruciformis* the second meiotic division is not synchronous. The upper dyad cell is at prophase whereas the lower dyad cell is observed at metaphase (Febulaus and Pullaiah, 1998).

In some members the upper dyad cell either fails to divide or even if divides, a wall is not formed resulting in a triad. Frequent occurrence of triads has been reported in *Poa pratensis* (Tinney, 1940), *Panicum maximum* (Warmke, 1954), *Pennisetum ciliare* (Snyder *et al.*, 1955), *Hierochloe odorata* (Norstog, 1963), *Cymbopogon martinii* (Bhanwra *et al.*, 1982; Bhanwra, 1988), *Arundinella nepalensis, Lolium perenne* and *Glyceria tonglensis* (Bhanwra, 1988).

In most grasses the isodiometric microphylar dyad cell is smaller than the chalazal member. The timing of the divisions of meiosis II is variable in Poaceae. In *Eustachys*, the division in the microphylar member always lags behind that of the chalazal one (Aulbach Smith and Herr, 1984). Such condition is also reported in *Festuca microstachys* (Maze and Bohm, 1977), *Oryzopsis hendersoni* (Mehlenbacker, 1970) and *Agrostis pilosula* (Muniyamma, 1976).

Callose, a β-1, 3-glucan, is synthesized and deposited in the walls of megaspore mother cells of angiosperms that produce either normal monosporic or bisporic embryo sacs. Such deposits are absent in angiosperms that produce tetrasporic embryo sacs and were recently shown to be absent in diplosporous *Elymus rectisetus* (Carman *et al.*, 1991) *Tripsacum* spp. (Leblanc *et al.*, 1995) and *Poa nemoralis* (Naumova *et al.*, 1993). Peel *et al.* (1997) studied about the megasporocyte callose in apomictic buffel grass, kentucky blue glass, *Pennisetum squammulatum* and weeping love grass.

Presence and distribution of callose in cell walls during megasporogenesis was described by Schwab (1991) in *Diarrhena.* Such deposition of callose in the walls during megasporogenesis is also reported by Maze and Bohm (1974) in *Agrostis interrupta.*

A thin layer of callose surrounds the megasporocyte at prophase I of meisis. At the dyad stage there is callose between and the surrounding the two cells. In *Diarrhena mandshurica* the micropylar megaspore is almost obscured by the callose reaction (Schwab, 1971). Bennett (1981) reported the presence of dark-staining material in association with the developing megaspore of *Avena.* The presence of callose in these three Festucoid grasses shows that presence of callose is a diagnostic feature of the grasses.

In all the sexually reproducing members of the Gramineae whether it is a tetrad or triad, the chalazal megaspore is functional and the development of embryo sac follows the monosporic Polygonum type (Figs. 10D–K, 11G–K). However, in *Poa pratensis* Anderson (1927) reported that the micropylar megaspore of a linear tetrad is occasionally functional. In the same species Grazi *et al.* (1961) reported that all the four of the megaspores degenerate. Stover (1937) also reported that in *Poa* micropylar megaspore of the tetrad is functional. In *Echinochloa frumentacea* (Narayanaswami, 1955b) occassionally three megaspores of a tetrad are functional and upper most megaspore of the tetrad is reported to get degenerated.

The degeneration pattern in non-functional megaspores is a constant feature in Poaceae. In *Eustachys* (Aulbach-Smith and Herr, 1984) the two micropylar megaspores degenerate first soon after their formation. The third megaspore degenerates later probably as a consequence of enlargement of the functional chalazal metgaspore. This pattern is also reported in some species of *Eleusine* (Narayanaswami, 1955c; Chandra 1963b), *Festuca* (Maze and Bohm, 1977) and *Eragrostis* (Mengesha and Guard, 1966). In *Eragrostis unioloides* (Untawale *et al.*, 1969) the second megaspore degenerates first. In *Stipa tortilis* and *Oryzopsis miliacea* (Maze *et al.*, 1970) the third megapsore degenerates first. In *Oryzopsis kingii* the three micropylar megaspores sometimes do not undergo degeneration until the two-nucleate female gametophyte stage (Kam and Maze, 1974). Cass and Peteya (1985) reported that in *Hordeum vulgare* the sequence of degeneration of the three micropylar megaspores vary. The degeneration begins in the central megaspore of the micropylar triad and followed in the megaspore adjacent to the functional megaspore.

In sexually reproducing species of Poaceae the chalazal megaspore is functional and the development of embryo follows monosporic Polygonum type. However only one exception to this was reported by Mohamed and Gould (1966) where wall formation will not take place during nuclear division in the megaspore mother cell and due to this coenomegaspore is formed. Hence in *Bouteloua curtipendula* they reported tetrasporic Adoxa type of embryo sac development. Emery and Guy (1979) are of the opinion that the delayed cytokinesis in *Zizania aquatica* until after meiosis II possibly indicate a trend towards tetrasporic development.

The development of female gametophyte in apomictic species differs from the normal development. In apomictic species degeneration of megaspre mother cell is reported in *Dicanthium pseudoischaemum*, *Cenchrus ciliaris* (Febulaus and Pullaiah, 1995, 1998), *Heteropogon contortus* (Gupta, 1968; Tethil, 1968) etc. Degeneration of sexual mechanism at dyad or tetrad stages has been reported in *Paspalum secans* (Snyder 1957), *Capillipedium parviflorum* (Bhanwra *et al.*, 1982; Bhanwra, 1988), *Pennisetum orientale* (Bhanwra, 1988) and also in *Panicum notatum* (Febulaus, 1992). In all these species the four megaspores of a tetrad degenerate and all these species were found to be apomictic.

Emery and Brown (1957) reported the formation of extra ovular embryo sac in *Digitaria valide* and *Fingerhuthia africana*. In the former species occasionally egg apparatus and part of embryo sac is observed to penetrate broad ovular attachment and come outside the integuments, while in *F. africana* it is reported to grow out through the micropylar end into a special cavity at the base of ovary. The micropyle formation here is delayed and nucellar epidermis disintegrates. In such instance the authors believe at least early development of embryo must be completed outside the ovule.

Development of embryo sac in apomictic members has been described in the Chapter Apomixis.

CHAPTER 4

FEMALE GAMETOPHYTE

In Poaceae the female gametophyte, in sexually reproducing members, is typically 7-celled structure consisting of large central cell with two polar nuclei which later forms the secondary nucleus, an egg apparatus with two syergids and an egg cell, and three antipodals. The development of embryo sac in all the sexually reproducing members of the family Poaceae is of Monosporic Polygonum type. However only one exception to this normal type was made by Mohamed and Gould (1966) in *Bouteloua curtipendula* complex. In this species the development of embryo sac is reported as Adoxa type. In majority of species the development of embryo sac is reported as monosporic Polygonum type.

MONOSPORIC POLYGONUM TYPE

In majority of Poaceae whether it is a linear or T-shaped tetrad or triad, the chalazal megaspore of the tetrad is functional and the embryo sac is formed from this chalazal megaspore. After the non-functional megaspores start to degenerate, the functional megaspore starts to enlarge. During early development megagametophyte is more or less oblong. But as the development continues, the upper side of the megagametophyte begins to grow upward (Maze and Bohm, 1974). The enlargement of the functional megaspore involves both an increase in the amount of cytoplasm and sharp increase in vacuolation. A large vacuole occupies the central part of the embryo sac.

Increase in the functional megaspore is accompanied by degeneration of nucellar epidermis. Its individual cells collapse and the cell walls remain intact. Their nuclei become very elongated and structureless.

The functional megaspore contains dense, vacuolate to highly vacuolate cytoplasm. Its nucleus is centrally located. The functional megaspore undergoes rapid enlargement and functions as embryo sac initial. This develops into 1-nucleate embryo sac. The nucleus of the functional megaspore undergoes first mitotic division resulting in the formation of 2-nucleate embryo sac (Fig. 11H). Due to the appearance of the large central vacuole, the two nuclei are pushed towards the opposite poles of the central cell (Fig. 11G). The two nuclei undergo second mitotic division forming two nuclei at each pole (Figs. 9J,

10H, 11I). The large central vacuole separates the embryo sac cytoplasm into two main areas, one at the micropylar end and the other at the chalazal end. The two nuclei at each pole undergo third mitotic division resulting in the formation of four nuclei at each pole. This 8-nucleate stage of unorganised embryo sac is of short duration. Immediately after the third mitosis, cytoplasmic cleavage takes place resulting in the formation of three uninucleate cells at each pole. Of the four nuclei at the micropylar end, three organise into egg apparatus and the fourth one functions as upper polar nucleus. The nuclei of the chalazal quartet form three antipodals and the fourth nucleus functions as lower polar nucleus. The organisation of the mature embryo sac in sexually reproducing species is 8-nucleate and seven-celled structure having two synergids, one egg cell, two polar nuclei and three antipodals (Fig. 11J). Thus the development of organised embryo sac follows Monosporic Polygonum type of Maheshwari (1950).

TETRASPORIC ADOXA TYPE

Mohamed and Gould (1966) reported Adoxa type of embryo sac development in *Bouteloua*. The archesporial cell differentiates deep in the nucellus before the integuments become fully developed. This archesporial cells enlarges and functions directly as megaspore mother cell. This undergoes meiotic divisions forming four-nucleate megagametophyte without any indication of a linear arrangement of the nuclei or separation by cell walls. Two nuclei are present at the micropylar end and two nuclei are present at the chalazal end. These four nuclei undergo mitotic division resulting in the formation of 8-nucleate embryo sac which develops into a mature megagametophyte consisting of an egg, two synergides, two polar nuclei and three antipodals. Thus the development of embryosac is of "Adoxa type" in which all 4 megaspore nuclei undergo one more mitotic division to form an eight nucleate embryo sac.

This type of embryo sac development was reported in *Bouteloa warnockii, B. media, B. uniflora* var. *uniflora, B. uniflora* var. *coahuilensis, B. curtipendula* var. *curtipendula* (Mohamed and Gould, 1966).

In *Bouteloua* spp. some mature embryo sacs examined were 3-nucleate and appears to lack synergids and antipodals (Mohamed and Gould, 1966). Bashaw and Holt (1958) also reported such cases in *Paspalum dilatatum*. The three-nucleate embryo sac is formed when one of the two nuclei in the binucleate stage fails to divide and the other divides only once.

ORGNISATION OF THE EMBRYO SAC

The organised megagametophyte, which is spindle shaped, oblong or bean-shaped, is either oriented parallel to the long axis of the ovule or as in Bambusoid, Arundinoid and Pooid grasses shows slight curvature towards the chalaza. In most of the Chloridoid grasses the megagametophyte shows a prominent curvature while in panicoid grasses it is placed almost parallel to the longitudinal axis of the ovule (Bhanwra, 1988).

After the last nuclear division in the female gametophyte the cells of the embryo sac are formed within the wall of the parent megaspore. In maize the wall of the embryo sac is multilayered and pectocellulosic (Chebotaru, 1975). According to Maze and Lin (1975) the wall surrounding the megagametophyte consists of three portions. The inner most portion is most likely the megagametophyte wall, the outermost portion is the walls of the nucellar cells and the middle is the remnants of the megaspore wall or the remains of nucellar cells that were crushed during megagametophyte development. The final organisation of the embryo sac in majority of Poaceae is 7-celled structure consisting of an egg, two synergids, two polar nuclei and three antipodals. The number and ploidy of the antipodals is variable.

Jane (1997) gave the details of ultrastructure of maturing egg apparatus in *Arundo formosa.* All three cells of the newly formed egg apparatus surrounded by incomplete cell walls initially have similar cellular contents. However, as the egg apparatus develops, the egg and the synergids gradually differentiate in cellular contents prior to the initiation of filiform apparatus. The egg develops many starch containing plastids. The synergids become rich in rough endoplasmic reticulum (ER) and dictyosomes. These differences are retained through out maturation. The micropylar common wall of the two synergids produces many cell-wall ingrowths, which constitute the filiform apparatus. Endoplasmic reticulum and dictyosomes are involved in filiform — apparatus formation. During maturation the egg shows some changes: the nucleus migrates towards the cell centre; the degree of vacuolation and the number of starch grains and oil bodies increase; the amount of ER and dictyosomes decrease. During synergid maturation, the organelles gradually show a polarized distribution; the plastids are present only at the chalazal pole, the most mitochondria and dictyosomes are distributed at the micropylar pole. After pollination, vacuolization usually occurs at the chalazal pole of the persistent synergid, and the degenerated synergid becomes filled with electron-dense materials. During egg apparatus expansion cell wall materials are deposited and condense at the chalazal pole, forming electron-dense particles at maturity. Plasmodesmata are always present in the micropylar gametophytic cell walls of the egg apparatus (Jane, 1997).

Egg Apparatus

Synergids

The synergids are elongated cells present at the micropylar end of the embryo sac. The synergids are pointed or hooked or pyriform or pear-shaped. The synergids in most grasses are ephermeral (Narayanaswami, 1953; Aulbach-Smith and Herr, 1984; Bhanwra, 1988). However in *Zea mays* (Randolph,

1936; Cooper, 1937) and *Euchlaena mexicana* (Cooper, 1937) they remain persistent. In *Triticum* the synergids persist till 4- or 6-celled embryo stage. In *Festuca microstachys* (Maze and Bohm, 1977) synergids are seen where embryo development has started.

Synergids in majority of Gramineae are usually pyriform or subhemispherical cells, adjacent to the micropyle as reported in *Arundinaria* (Yashida, 1963), *Zea* (Diboll and Larson, 1966), *Hordeum* (Cass and Jensen, 1970), *Oryzopsis* (Mehlenbecher, 1970), *Stipa* (Maze and Bohm, 1973), *Agrostis* (Maze and Bohm, 1974), *Festuca* (Maze and Bohm, 1977) and *Axonopus* (Anton, 1982).

The wall of the synergid is incomplete. A distinct wall is present at the micropylar one third of the cell and it gradually thins towards the chalazal end. The chalazal one third of the cell lacks a definite wall and the protoplast of the synergid is separated from the central cell by plasmamembranes.

At the micropylar end of each synergid a prominent structure called filiform apparatus is present. This filiform apparatus is a highly convoluted extension of the micropylar portion of the synergid wall. The presence of abundance starch grains in a young synergid and their virtual absence from a fully developed synergid led Yu and Chao (1978) to suggest that in *Paspalum* the filiform apparatus seems to be formed mainly from substances transformed from starch grains. The filiform apparatus serves to increase the surface area of the plasma membrane in the micropylar portion of the synergids (Maze and Lin, 1975). In the degenerated synergid the function of FA is the control of pollen tube growth (Maze and Lin, 1985). In the persistent synergid the function of FA has been hypothesised on the transfer of material into the megagametophyte (Gunning and Pate, 1969). In *Stipa elmeri* the persistent synergid with its polarity may aid in establishing the polarity of the egg (Maze and Lin, 1975).

The cytoplasm of the synergid is highly polarised. The chalazal region of the cell is occupied by a large vacuole and large amount of cytoplasm and nucleus is present towards the micropylar region of the cell. The cytoplasm of the synergid is rich in cell organanels. Mitochondira, dictyosomes are concentrated in the micropylar half of the synergid near the filiform apparatus. An abundance of small vesicles are also present in this region. Profiles of endoplasmic reticulum with associated ribosomes are oriented more or less parallel to the axis of the cytoplasmic polarity in the middle region of the cell. Plastids with few internal lamellae are clustered between the nucleus and the vacuoles at the chalazal end of the cell (Diboll, 1968b).

Synergids with haustorial projections were of rare occurrence in Poaceae. However extensive synergid haustoria have been reported by Philipson and Connor (1984) in some Danthonioid grasses (Fig. 12). In *Cortaderia* (Philipson, 1981) finger-like projections arise from the filiform apparatus and penetrates the micropyle. The entire inner surface of the haustorium is provided with wall projections. These are involved in the absorbtion and conduction of nutrients to the synergids. The tips of the synergids grow beyond the micropyle in

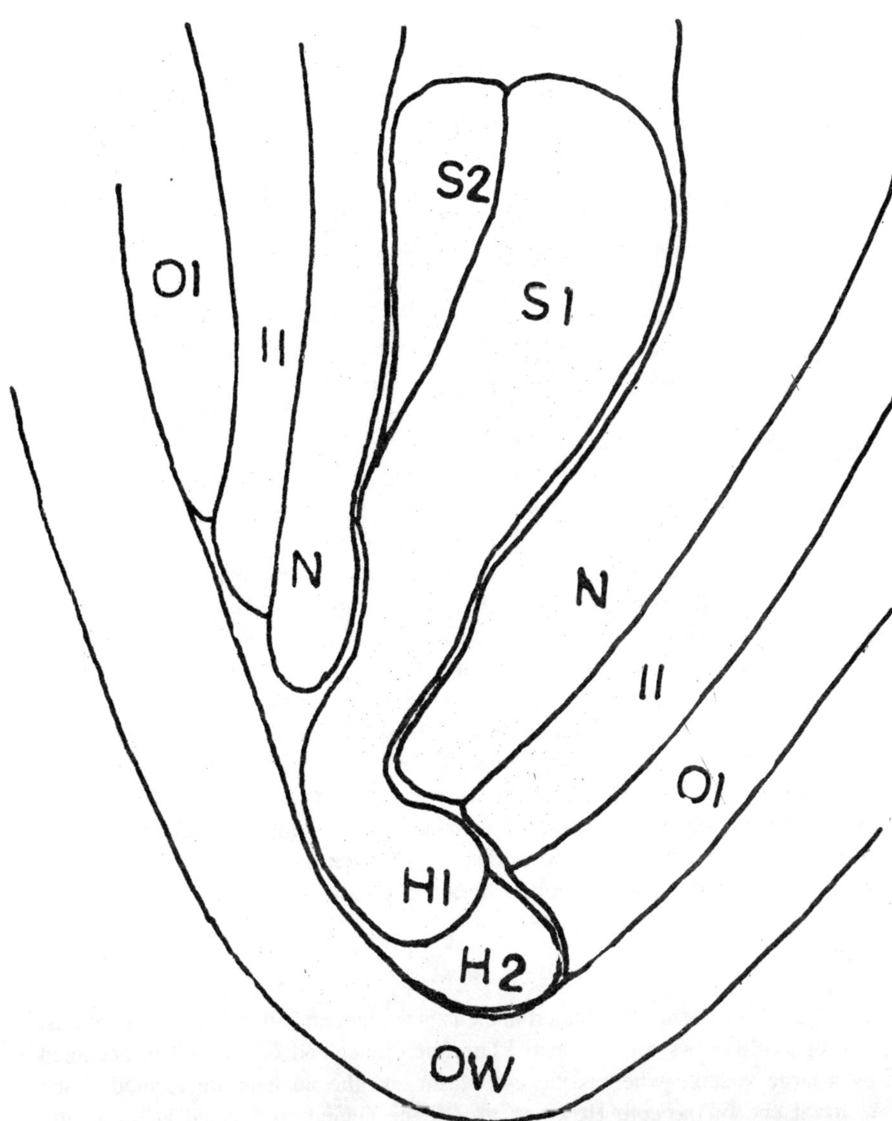

Fig. 12. Micropylar end of the ovule of *Danthonia spicata.* Two synergids (S_1 and S_2) with haustorial extensions (H_1 and H_2) project through the micropyle and grow in the same direction along the ovary wall (OW) and against inner (II) and outer integuments (OI). N = Nucellus. (Philipson & Connor, 1984).

Cortaderia selloana (Philipson, 1977) the haustoria extend in opposite direction between the outer integument and the ovary wall. In *C. jubata* the haustorium penetrated the nucellus laterally extending into the integuments. Later it reached the ovary wall adjacent to the placenta. This haustorium persists up to the early globular stage of the proembryo (Philipson, 1978a).

In *Chionochloa, Cortaderia, Danthonia* (Fig. 12), *Erythranthera, Lamprothyrsus, Rytidosperma* and *Sieglingia* the synergids extends through the micropyle to lie between the ovule and ovary wall. Extensions through the micropyle are massive and wide. They reach the outside of the ovule in the micropylar region by penetrating integumentary cells close to the micropyle (Philipson and Connor, 1984).

Verboom *et al.* (1994) has recently made extensive studies on the haustorial synergids in Danthoid grasses. They reported well developed synergid haustoria protruding through micropyle in *Merxmuellera rufa, M. disticha, M. dura, M. stricta, Pentaschistis chippendalliae, P. eriostoma, P. patula, P. velutina, Tribolium obliterum, Urochlaena pusilla, Danthonia pallida, Chaetobromus dregeanus, C. involucratus, Pseudopentameris, Rytidosperma erianthum, R. clelandii, R. vickeryi, Karroochloa purpurea, K. tenella, Cortaderia selloana* and *Erythranthera australis.*

Philipson and Connor (1984) suggested that the absence of haustorial synergids in nonarundinoid grasses is indicative of their derived (apomorphic) nature, and that they might therefore be of some use in revealing monophyletic group within the arundinoids. Significantly, this character appears to be restricted to members of the tribe Danthonieae, a large group of genera with an Austral center of distribution.

The Synergids are thought to function in the production of substances which attract the pollen tube to the megagametophyte (Van der Pluijm, 1964). In some species they may also participate in nourishing the young embryo (Jensen, 1965a). Chao (1971) postulated that in *Paspalum* the synergids are involved in enzyme production and secretion. These enzymes are presumed to be responsible for the degeneration of integumentary cells leading to the release of PAS positive substances into micropyle. In general opinion the degenerating synergid forms the seat for pollen discharge in the embryo sac.

Egg

The egg with its centrally located nucleus is the largest cell of the egg apparatus. The egg cell shows high polarity. The micropylar end of the cell is occupied by a large vacuole whereas the cytoplasm and the nucleus are located at the chalazal end of the cell. However in *Zea* the nucleus is located at the centre and many small vacuoles are present along the periphery of the cell (Bhojwani and Bhatnagar, 1974). In *Stipa elmeri* (Maze and Lin, 1975) the egg cell is also vacuolate.

The young egg cell is rich in organelles such as plastids, mitrochondira and dictyosmes. The plastids contain starch and have poorly organised internal membranes. The egg mitochondria vary from round to elongated and have a fairly well developed internal membrane system. Endoplasmic reticulum in some egg cells appear as small vesicles, in some, the smooth endoplasmic reticulum appear as flattened sheets around the nucleus, and in some egg cells the rough endoplasmic reticulam appears to be in sheets around the nucleus. A few dictyosomes with two or four stacks of lamellae are present in the egg cytoplasm.

Development of haustorial structures from the egg cell is very rare phenomenon in Poaceae. However Philipson (1981) reported that in *Cortaderia jubata* the egg cell rarely develops a haustorium.

Antipodals

Antipodal cells in the family Poaceae have attracted considerable attention with regard to their number, size, persistance and possible role in nutrition of young developing seed (Brink and Cooper, 1944; Narayanaswami, 1953). Cannon as far back as 1900 observed that in *Avena fatua* the antipodals increased in their number up to fertilization and begin to degenerate during the process of endosperm formation. The number of antipodals is highly variable. Within the family Poaceae their number may range from three to three hundred. The highest number of antipodals in the whole of angiosperms is met within *Sasa paniculata* (Yamaura, 1933) where 399 antipodals have been reported. Hoshikawa *et al.* (1960) observed that the three initial antipodals divide to produce about 40 cells in wheat. In very few species the antipodals remain three whereas in majority of the members they undergo divisions resulting in variable number.

In species with proliferating antipodals, the cells can develop with more than one nucleus. Once the antipodals begin to proliferate they appear somewhat stypical. Vacuoles are lacking in most, nuclei are lacking in many and many nuclei that present lack nucleoli. In *Saccharum benghalense* (Bhanwra, 1988) up to six nuclei have been reported. In *Pennisetum typhoideum* (Narayanaswami, 1953) the antipodals are coenocytic containing 3 to 8 nuclei. Khosla (1946) reported as many as 23 nuclei in each antipodal cell of *Pennisetum typhoideum.* Coenocytic antipodal cells are also reported in *Echinochloa frumentacea* (Narayanaswami, 1955b) and *Andropogon pumilus* (Kulkarni, 1982). In *Echinochloa colonum* and *E. crusgalli* the antipodals show an increase of nuclei up to 4 (Bhanwra and Choda, 1986). In *Zea* mays during additional divisions in the antipodals the walls of many cells remain incomplete, leaving protoplasmic continuities between adjacent cells. This results in the formation of multinucleate protoplasm or syncytium (Diboll and Larson, 1966).

In *Phleum boehmeri* (Joachimiak, 1981), the number of antipodals in an embryo sac varies considerably as well as their ploidy-level. Thirty-eight per cent of mature embryo sacs had three antipodals each, but the number increases to ten at the cellular stage of endosperm. The nucleus of the antipodal cell may be spherical, dumb-bell shaped or irregular. The restitution nucleus (after anaphase) is constricted in the central region. They undergo further inhibited divisions. The polyploid nuclei often show regular shape, and rhythmical growth in volume. With ploidy-level the volume increases in geometrical proporation: n = 70–80 μm^3, 2n = 151–173 μm^3, 4n = 315–376 μm^3, 8n = 681–777 μm^3, 16n = 1121–1452 μm^3, 32n = 2073–3076 μm^3, 64n = 4820–5060 μm^3, 128n = 10685–12075 μm^3 and 256n = 20093–20420 μm^2. The nuclei with irregular shape have many nuclei. The highly polyploid antipodal cell may become multinucleate. In one case out of the four nuclei two were 16n, one 32n and one 256n. They degenerate at the multicellular stage of the embryo. The degeneration is marked by vacuolation of cytoplasm, collapse of cell walls, and agglutination of chromation or disintegration of nuclei (Joachimiak, 1981).

Antipodal haustoria are reported in *Agrostis pratensis, Festuca pratensis, Holcus lanatus* (Terzijski and Kuriston, 1973).

Ultrastructural or histochemical aspects of antipodal cells are not available. According to Diboll and Larson (1966) the antipodal cells in maize have abundant mitochondria, plastids and multicisternal dictyosomes. The cytoplasm is full of endoplasmic reticulum or the dictyosomes. Frequently the antipodal cells bordering the nucellus show finger-like wall projections. In *Stipa* (Maze and Lin, 1975) and *Zea* (Diboll, 1968) the presence of these transfer wall-like projections are accompanied by the presence or very active cytoplasm. The cytoplasm contains abundant organelles such as mitochondria, dictyosomes, plastids and ribosomes. The endoplasmic reticulum is extensive and consists of parallel and partly concentric cisternae.

Antipodals are usually ephemeral and they become degenerated after fertilization. However persistent antipodals have been reported in *Zea mays* (Randolph, 1936; Cooper, 1937), *Euchlaena mexicana* (Cooper, 1937), *Pennisetum typhoideum* (Krishnaswami and Ayyangar, 1941; Narayanaswami, 193) and also in *Chionachne koenigii* (Satyamurty, 1984b). The antipodal cells remain persistent up to 2-celled proembryo stage in *Stipa lemmonii* (Maze *et al.*, 1972) and up to 2–4 celled stage of embryo in *Dendrocalamus hamiltonii* (Harigopal and Manasiram, 1981). The antipodal cells persist till cellularization of the endosperm in *Triticum* (Bhatnagar and Chandra, 1976).

In *Thyrsostachys oliveri, Leptochloa chinensis, Chloris barbata* degeneration of antipodals take place at 2-celled proembryo stage. In *Polypogon fugax, P. monspeliensis* and *Tragus roxburghii* antipodals degenerate at 4-celled stage and in *Phalaris minor* and *Setaria glauca* at globular stages (Bhanwra, 1988).

Within the family Poaceae, the antipodal cells exhibit great variation in respect to their position in the mature embryo sac. Shadowsky (1926) gave a list of several grasses where position of antipodals in the embryo sac was either chalazal or lateral. He reported that with a few exceptions, majority of the grasses belonging to the Pooideae the antipodals are lateral in position (Fig. 13) while those of the Panicoideae are chalazal in position. A similar observation

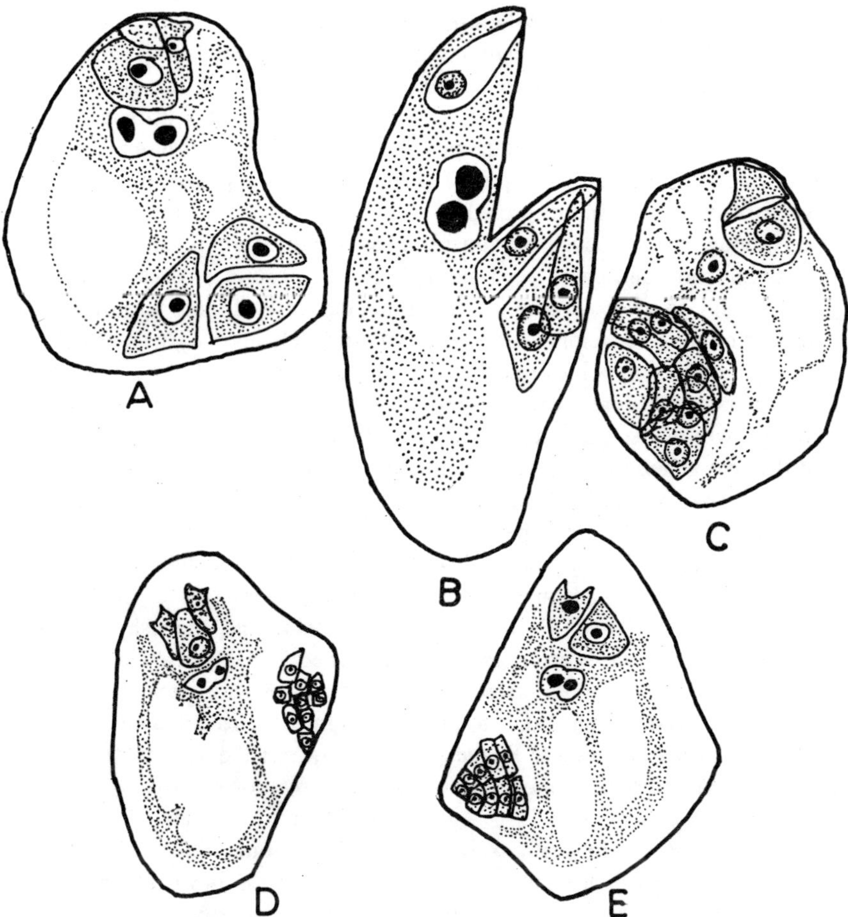

Fig. 13. Lateral position of antipodals in Pooideae.A. *Tragus biflorus* (Chandra, 1975), B. *Cbloris roxhurghiana* (Febulaus and Pullaiah, 1991a), C. *Dactyloctenium aegyptium* (Chandra, 1963), D. *Egagrostis unioloides* (Chandra, 1976, E. *Eragrostis coarctata* (Chandra, 1975).

was made by Chandra (1963b), Venkateswarlu and Devi (1964), Bhanwra (1988) and Febulaus and Pullaiah (1991a,b, 1992a,b, 1993a,b, 1996).

Definite role has been attributed to the antipodals. In *Eleusine coracana* (Narayanaswami, 1955) the antipodals appear to be of glandular or secretory type acting as reservoirs of food materials. Occasionally they have pointed ends and they may also have haustorial function. Rarely they become aggressive.

The presence of transfer cell-like walls suggests that they are involved in the transfer of nutrients and serve as a pathway for the metabolites from the nucellus to the central cell. They serve as storage centres for the developing endosperm and embryo. Sometimes the antipodal cells regulate the development of the adjacent endosperm (Willemse and Van Went, 1984). In *Triticum vulgare* (Guenther, 1927) and *T. aestivum* because of their intermediate position in the nutrient stream, Bhatnagar and Chandra (1975) ascribed the role of conduction and nutrition to these antipodal cells. Maze and Lin (1975) attributed that the grass antipodals are involved in the production of growth controlling substances. According to Brink and Cooper (1944) the antipodals are involved in the control of early endosperm growth.

Although the antipodals are considered to be nutritive in function, sometimes they disfunction and cause seed sterility. In intergeneric crosses of *Hordeum jubatum* □ □ *Secale cereale* □, the antipodal cells do not proliferate, the endosperm fails to develop, and the embryo ultimately aborts. Brink and Cooper (1944), therefore, hypothesized that the antipodals behaved abnormally under the influence of a foreign pollen, fail to secrete substances necessary for the initial growth of the endosperm, and result in irregularities of development, so that the intergeneric hybrid fails. Similarly, in a cross of *Elymus virginicus* □ and *Agropyron repens* □, the antipodals to invoke the growth of the endosperm (Beaudry, 1951). The endosperm nuclei become irregular in shape and size after 48 hr of pollination and result in the eventual degradation of the hybrid.

The role of antipodals in interspecific and intergeneric crosses in Gramineae has been much debated. Brink and Cooper (1944) and Maheshwari (1950) attributed a secretary role and pointed out that the antipodals nurse the endosperm during earlier stages of its development. Contrarily, Thompson and Johnson (1945) considered the antipodals to play an insignificant role in a cross between barley and rice. According to Beaudry (1951) however the activity of antipodal cells alters the nutrient supply causing the breakdown of endosperm in a cross between *Elymus virginicus* L. □ *Agropyron repens* (L.) Beauv. During post-fertilisation stages, in *Triticum aestivum* (Morrison, 1955) the antipodal complement neither divides, nor increases in size, whereas they get hypertrophied in *T. polonicum, T. spelta, T. aegilopoides* and *T. spelta* female □ *T. polonicum* (Wakakuwa, 1934). Chandra (1970) also reported increase in the number and volume of antipodal cells in *T. aestivum.*

Central Cell

Central cell is the largest cell of the embryo sac. The central cell is highly vacuolate. Its cytoplasm is restricted to its periphery (Maze and Bohm, 1973). The polar nuclei lie in the centre of the central cell when the embryo sac is seven-celled. The two polar nuclei fuse prior to fertilization. In *Spinifex littoreus* they do not fuse until fertilization (Jayalakshmi and Lakshmanan, 1982). In *Agrostis interrupta* (Maze and Bohm, 1974) the polar nuclei fuse before the egg apparatus is formed.

The central cell cytoplasm contains long profiles of rough endoplasmic reticulum. In the cytoplasm around the egg and also around the polar nuclei the RER present mostly as vesicles. A large number of mitochondria and amyloplasts are concentrated around the polar nuclei. In addition to these plastids, dictyosomes and lipid bodies are also found in the central cell cytoplasm. The central cell plastids are spherical with simple internal membrane system. Dictyosomes are mostly consisted of four stacks of lamellae (Maze and Lin, 1975).

The cytoplasm of the central cell is generally believed to nourish the egg during its development (Diboll, 1968).

Ultrastructure of Apomictic Embryo Sacs

Ultrastructural studies on sexual embryo sacs in Angiosperms are widespread, but hardly any information is available at this level for asexual (apomictic) embryo sacs. Sexual and asexual embryo sacs of *Pennisetum* were studied in detail by Chapman and Busri (1994) to reveal similarities and/or differences in cell number, their cytoplasmic content, in the distribution of wall projections and embryo sac boundaries. At this level, the two types of embryo sacs are remarkably similar and apart from small differences in wall structure, this confirms the view that the chief contrasts concern nuclear number and function.

TABLE II. Structure and Development of ovule and Embryo Sac

Species investigated and Author	Ovule	Arche-sporium	Megaspore tetrad	Type of embro sac develop-ment	Synergids	Antipodals
Acrachne racemosa (Bhanwra, 1988)	Hemiana-tropous, bitegmic and tenui-nucellate	Single-celled	Linear, chalazal funcitonal	Polygonum type	Hooked	3–9 antipodals
Acroceras munroanum (2n = 18) (Shobha and Sindhe, 1992)	Hemiana-tropous bitegmic and crassi-nucellate	"	"	"	—	8 anti-podals uni-nucleate
Agrostis interrupta (Maze and Bohm, 1974)	Hemiana-tropous, bitegmic and tenui-nucellate	"	T-shaped tetrad, chalazal functional	"	Hooked	22 anti-podals, 3-4-nucleate, some multi-nucleate
A. pilosula (Muniyamma, 1976)	"	"	"	"	—	—
Alloteropsis cimicina (Venkateswarlu and Devi, 1964)	Anatropous bitegmic and pseudocrassi-nucellate	"	Chalazal functional	"	—	3–6 anti-podals
Andropogon pumilus (Kulkarni, 1982)	Hemiana-tropous, bitegmic and tenui-nucellate	"	Linear, chalazal functional	"	—	3 anti-podals, coenocytic
Apluda mutica (Chandra, 1963b; Bhanwra and Pathak, 1987)	Hemiana-tropous, bitegmic and pseudocrassi-nucellate	"	Linear or T-shaped, chalazal functional	"	—	3-5 cells
Aristida adscensionsis (Bhanwra *et al.*, 1982; Bhanwra, 1988)	"	"	"	"	Hooked	18 cells

Species investigated and Author	Ovule	Arche-sporium	Megaspore tetrad	Type of embro sac development	Synergids	Antipodals
Aristida funiculata (Febulaus and Pullaiah, 1993b)	Hemiana-tropous bitegmic and tenui-nucellate	Single-celled	Linear or T-shaped, chalazal functional	Polygonum type	Hooked	6–13 cells
Aristida hystrix (Febulaus and Pullaiah, 1993b)	Hemiana-tropous bitegmic and pseidocrassi-nucellate	Single-celled	Linear or T-shaped, chalazal functional	Polygonum type	Pear-shaped	6–13 cells
Aristida mutabilis (Febulaus and Pullaiah, 1993b)	"	"	"	"	Hooked	3–6 cells
Aristida setacea (Febulaus and Pullaiah, 1993b)	Hemiana-tropous, bitegemic and tenui-nucellate	"	"	"	Hooked	3–9 cells
Arundinella mesophylla (Basappa and Muniyamma, 1981)	—	"	Linear, chalazal functional	Polygonum type	No hooks De-generate prior to ferti-lization	2 only.
A. nepalensis (Bhanwra *et al.*, 1982; Bhanwra, 1988)	Hemiana-tropous bitegemic and pseudocrassi-nucellate	"	"	"	—	About 18 cells
A. purpurea (Basappa and Muniyamma, 1981)	—	"	"	"	—	Coenocytic antipodal complex
Arundo donax (Bhanwra *et al.*, 1982; Bhanwra, 1988)	Hemiana-tropous bitegemic and pseudocrassi-nucellate	Single-celled	Linear or T-shaped	Polygonum type	Hooked	36 uni-nucleate antipodals

(*contd.*)

(*Table II contd.*)

Species investigated and Author	Ovule	Arche-sporium	Megaspore tetrad	Type of embro sac develop-ment	Synergids	Antipodals
Avena fatua (Cannon, 1900)	—	—	—	—	—	Antipodals, increase in number
Axonopus arenosus (Gledhill, 1967)	Bitegmic	Single-celled	T-shaped tetrad, chalazal functional	Polygonum type shows automixis	—	Antipodal increase in number
Axonopus brevipedu-ncularis (Gledhill, 1967)	"	"	"	"	—	"
Axonopus compressus (Gledhill, 1967)	"	"	"	Polygonum type	—	12 antipodals
Axonopus flexuosus (Gledhill, 1967)	"	"	"	Polygonum type shows automixis	—	Antipodals incrase in number
Bambusa tulda (Harigopal and Mohan Ram, 1987; Bhanwra, 1988)	Hemiana-tropous bitegemic and pseudocrassi-nucellate	Single-celled	Linear, chalazal functional	Polygonum type	Hooked	12 anti-podals
Bothriochioa odorata (Bhanwra, 1988)	Hemianatrop -ous, bitegemic and pseudo-crasssi-nucellate	Single-celled	Linear, all megaspores degenerate	Monopolar, 4-nucleate unreduced embryo sac	—	4–6 antipodals
Bouteloua curtipendula (Mohammed and Gould, 1966)	—	—	Coenomega-spore	Tetrasporic Adoxa type	—	—
Brachiaria distachya (Bhanwra *et al.*, 1982; Bhanwra, 1988)	Hemiana-tropous, bitegmic and pseudo-crassinu-cellate	single celled	Linear, chalazal functional	Polygonum type	—	5–8 antipodals

Species investigated and Author	Ovule	Arche-sporium	Megaspore tetrad	Type of embro sac develop-ment	Synergids	Antipodals
B. eruciformis (Febulaus and Pullaiah, 1998)	Hemianatrop -ous, bitegmic and pseudo crassinu-cellate	Single celled	Linear, chalazal functional	Polygonum type	Hooked	5–8 antipodals
B. ramosa (Chandra, 1963b; Bhanwra *et al.*, 1985)	"	"	Linear or T-shaped, chalazal functional	"	Pear shaped	5–7 antipodals
B. reptans (Febulaus & Pullaiah, 1991b)	"	"	T-shaped, chalazal functional	"	"	4 antipodals
B. remota (Venkateswarlu and Devi, 1964)	Anatropous, bitegmic and pseudocrassi-nucellate	"	Linear or T-shaped, chalazal functional	"	"	"
Brachypodium sylvaticum (Bhanwra, 1988)	Hemiana-tropous bitegmic and pseudocrassi-nucellate	"	"	"	Hooked	3–8 antipodals
Bromus sp. (Beck & Horton, 1932)	— —	" "	T-shaped	Polygonum type	"	"
Bromus inermis (Nielsen, 1947a)	—	"	—	"	—	—
B. unioloides (Bhanwra, 1988)	Hemiana-tropous bitegmic and tenuinucellate	Single celled	Linear, chalazal functional	"	Hooked	12 antipodals
Calamogrostis perplexa (Greene, 1984)	Hemiana-tropous bitegmic and tenuinucellate	Single-celled	Linear or T-shaped, chalazal functional	Sexual Polygonum type	Hooked, filiform apparatus present antipodals	10 or more multi-nucleate

(*contd.*)

(*Table II contd.*)

Species investigated and Author	Ovule	Arche-sporium	Megaspore tetrad	Type of embro sac develop-ment	Synergids	Antipodals
C. pickeringii (Greene, 1984)	"	"	"	"	"	"
Calamogrostis porteri (Greene, 1984)	"	"	"	Sexual Polygonum type	Hooked, filiform apparatus present	10 or more multi-nucleate antipodals
C. stricta (Greene, 1984)	"	"	No meiosis	Diplosporous	"	"
Capillipedium huegelli (Choda *et al.*, 1980)	"	"	Linear tetrad, chalazal functional	Sexual Polygonum type Aposporous unreduced 4-nucleate embryo sacs	—	9 antipidals
C. parviflorum (Bhanwra *et al.*, 1982; Bhanwra, 1988)	Hemiana tropous, bitegmic and pseudo-crassi-nucelate	Single celled	Linear or T-shaped. All the megaspores degenerate	4-nucleate unreduced embryo sacs are formed from nucellar cells	—	—
Cenchrus biflorus (Bhanwra *et al.*, 1982; Das & Islam, 1977)	"	"	Linear, chalazal functional	Polygonum type	—	22–25 antipodal cells
C. ciliaris (Febulaus & Pullaiah, 1995)	"	"	No meiosis	Aposporic Panicum type	Hooked	3 antipodals
C. glaucus (Shanthamma, 1982)	Hemiana tropous, bitegmic	"	Linear tetrad	Sexual embryo sacs absent	—	—
Centropodia glauca (Verboom *et al.*, 1994)	Hemiana tropous, bitegmic and tenui-nucellate	—	—	Polygonum type	—	Antipodals proliferate
Chaetobromus dregeanus (Verboom *et al.*, 1994)	"	—	—	Polygonum type	Synergid haustoria well developed	Antipodals proliferate

Species investigated and Author	Ovule	Arche-sporium	Megaspore tetrad	Type of embro sac develop-ment	Synergids	Antipodals
C. involucratus (Verboom *et al.*, 1994)	"	—	--	"	"	"
Chionachne koenigii (Satyamurty, 1984b)	Bitegmic and pseudo-crassinuce-llate	1–3 cells	Linear tetrad, chalazal functional	Polygonum type	Pear-shaped	5–7 antipodal cells one- or two-nucleate
Chionochloa rigida (Maze and Bohm, 1974)	Hemiana-tropous, bitegmic and tenui-nucellate	—	—	—	Synergid haustoria well developed	Antipodals proliferate
Chloris barbata (Bhanwra *et al.*, 1981; Bhanwra, 1988)	"	Single celled	Linear chalazal functional	Polygonum type	Beaked	Antipodals increase in number
C. gayana (Chikkannaiah & Mahalingappa, 1976)	Camphylo-tropous. bitegmic and tenui-nucellate	1–3 cells	Linear, chalazal functional	Polygonum type	Beaked	Antipodals increase in number up to 12.
C. inflata (Chandra, 1963b)	"	—	—	"	—	—
C. roxburghiana (Febulaus and Pullaiah, 1991a)	Hemiana-tropous, bitegmic and tenuinucellate	Single celled	T-shaped tetrad, chalazal functional	"	Hooked	3 antipo dals, cells 3-nucleate
C. virgata (Strydom and Spies, 1994)	Anatrapous bitegmic	—	—	Polygonum type	—	—
Coix lachryma-jobi (Weatherwax, 1930)	Bitegmic	—	—	—	—	—
Cortaderia selloana (Philipson, 1977; Coastas-	Hemiana-tropous, bitegmic and tenui-nucellate	—	—	Polygonum type	Synergid haustoria well developed	Antipodals proliferate

(*contd.*)

(*Table II contd.*)

Species investigated and Author	Ovule	Arche-sporium	Megaspore tetrad	Type of embro sac develop-ment	Synergids	Antipodals
Lippman, 1979; Verboom *et al.*, 1994)						
C. jubata (Phillipson, 1977)		—	—	Egg may develop into haustorium	"	—
Cymbopogon caesius (Sreenivasa Rao, 1997)	Hemiana-tropous, bitegmic and pseudocrassi-nucellate	Single-celled	Linear chalazal functional	Polygonum type	Flask shaped	5–6 Anti-podals
Cymbopogon coloratus (Sreenivasa Rao, 1997)	"	"	"	"	"	"
Cymbopogon flexuosus (Sreenivasa Rao, 1997)	Hemiana-tropous, bitegmic and pseudo-crassi-nucellate	Single-celled	Linear chalazal functional	Polygonum type	Flask shaped	5–6 Anti-podals
Cymbopogon martinii (Brown and Emery, 1958; Choda *et al.*, 1982)	"	"	Linear chalazal functional	Polygonum type	—	12–15 antipodals
C. nardus (Choda *et al.*, 1982)	"	"	"	"	"	9–26 antipodals
Cynodon dactylon (Chandra, 1963b; Strydom and Spies, 1994)	Bitegmic, tenui-nucellate	"	"	"	—	—
Dactyloctenium aegyptium (Chandra, 1963a; Venkateswarlu & Devi, 1964)	Anatropous, bitegmic and tenui-nucellate	"	Chalazal functional	Polygonum type	—	6–10 antipodals

Species investigated and Author	Ovule	Archesporium	Megaspore tetrad	Type of embro sac development	Synergids	Antipodals
D. sindicum (Sharma *et al.*, 1981; Bhanwra, 1988)	"	"	Linear or T-shaped, chalazal functional	Polygonum type	—	12 cells, hypertrophied
Danthonia pallida (Verboom *et al.*, 1974)	Hemianatropous bitigmic and tenuinucellate	—	—	Apomictic	Synergid haustoria well developed	Antipodals proliferate
D. sericea (Philipson and Connor, 1984)	"	—	—	—	"	"
D. spicata (Philipson and Connor, 1984)	"	—	—	—	"	"
Dendrocalamus hamiltonii (Harigopal & Manasi Ram, 1981; Harigopal and Mohan Ram, 1987)	Hemianatropous, bitegmic and Pseudocrassinucellate	Single-celled	Linear or T-shaped, chalazal functional	Apomictic	Hooked	40–60, uninucleate, persists up to 2-4 celled stage of embryo
D. longispathus (Harigopal and Mohan Ram, 1987)	"	"	"	"	—	"
Desmostachya bipinnata (Bhanwra, 1986b; 1988)	Campylotropous, bitegmic and tenuinucellate	Single-celled	Linear or T-shaped, chalazal functional	Polygonum type	—	6 or 7 antipodals
Diarrhena americana (Schwab, 1971)	—	"	T-shaped, chalazal functional	"	—	—
D. mandshurica (Schwab, 1971)	—	—	Linear, chalazal functional	"	—	—

(*contd.*)

(*Table II contd.*)

Species investigated and Author	Ovule	Arche-sporium	Megaspore tetrad	Type of embro sac develop-ment	Synergids	Antipodals
Dicanthium annulatum (Reddy & D'Cruz, 1969a)	Bitegmic	Single-celled	—	Aposporous, 4, 8-nucleate embryo sac develop from nucellar cells	—	Antipodal complex
D. armatum (Reddy and D'Cruz, 1969c)	"	"	T-shaped or tetrahe-dral, chalazal functional	Polygonum type	—	10-15 antipodals each 1-5-nucleate
D. interme dium (Saran & de Wet, 1970)	"	"	Chalazal functional	Sexual Polygonum type Occassionally Aposporic 4-nucleate embryo sacs		60 cells, some are 2-4-nucleate
Digitaria adscendens (Kulkarni & Dnyansagar, 1981)	Tenui-nucellate	"	—	Polygonum type	—	—
Digitaria bicornis (Venkateswarlu & Devi, 1964; Febulaus and Pullaiah, 1996)	Hemiana-tropous bitegmic and pseudocrassi-nucellate	"	Linear,	Polygonum type	Pear-shaped	3-6 antipodals
D. biformis (Venkateswarlu & Devi, 1964)	Anatropous bitegmic and pseudocrassi-nucellate	"	"	"	Hooked	4 antipodals
D. ciliaris (Febulaus and Pullaiah, 1996)	Hemiana-tropous, bitegmio and pseudocrassi-nucellate	"	"	"	Hooked	4 antipodals
Dinebra retroflexa (Satyamurty, 1985b)	Campylo-tropus, bitegmic and tenui-nucellate	Single-celled	Linear tetrad, chalazal functional	Polygonum type	—	3 antipodals uni-nucleate

Species investigated and Author	Ovule	Arche-sporium	Megaspore tetrad	Type of embro sac develop-ment	Synergids	Antipodals
Diplachne fusca (Seshavatharam and Satyamurty, 1982)	"	—	Linear or T-shaped, chalazal functional	"	—	—
Dregeochloa pumila (Verboom *et al.*, 1994)	Hemiantro-pous, bitegmic and tenui-nucellate	—	—	Polygonum type	—	Antipodals proliferate
Echinochloa colonum (Bhanwra and Choda, 1986; Bhanwra, 1988)	Hemiana-tropous, bitegmic and pseudocrassi-nucellate	Single-celled	"	"	"	"
E. crusgalli (Bhanwra and Choda, 1986 Bhanwra, 1988)	Hemiana-tropous, bitegmic and pseudocrassi-nucellate	Single-celled	Linear or T-shaped, chalazal functional	Polygonum type	Hooked filiform apparatus present each	3 antipo-dals, each 2-4-nucleate
E. frumentacea (Narayanaswami, 1955b)	Anantropous, bitemgic and crassi-nucellate	"	T-shaped, chalazal functional, rarely 3 megaspores	"	Pyriform, hooked	3 antipo-dals, each 1-3-nucleate
Ehrharta pusilla (Verboom *et al.*, 1994)	Hemianatro-pous, Bitegmic and tenui-nucellate	—	—	Polygonum type	—	Antipodals proliferate
Eleusine africana (Chikkannaiah & Mahalingappa, 1975)	—	—	—	"	—	Antipodals increase in number
Eleusine compressa (Mahalingappa, 1977)	Hemianatro-pous, Bitegmic, tenui-nucellate	Multice-lluar	—	Polygonum type	—	3–16 antipodals
			—	3–16 antipodals		

(*contd.*)

(*Table II contd.*)

Species investigated and Author	Ovule	Arche-sporium	Megaspore tetrad	Type of embro sac develop-ment	Synergids	Antipodals
E. coracana (Khosla, 1946; Narayanaswami, 1955c)	Campylo-tropous, bitegmic	Single-celled	Linear tetrad, chalazal functional	Polygonum type	Pyriform	3 antipo dals rarely 4 to 5, uni-nucleate
E. indica (Chandra, 1963a; Bhanwra, 1988)	Hemiana-tropous, bitegmic and tenui-nucellate	"	"	"	Hooked	12 antipodals
Elymus rectisetus (Crane & Carman, 1987)	—	—	Sexual ones, Linear or T-shaped, chalazal functional	Both Sexual and apomictic, Sexual Polygonum type	—	"
Elymus scabrus (Crane & Carman, 1987)	—	—	Linear or T-shaped, chalazal functional	"	—	Antipodals proliferate
Elytrophorus spicata (Satyamurty, 1985a)	Campylo-tropous, bitegmic and tenui-nucellate	Single-celled	T-shaped chalazal functional	Polygonum type	—	3 antipo dals, uni-nucleate
Eragrostiella bifaria (Venkateswarlu & Devi, 1964; Febulaus & Pullaiah, 1992b)	Hemiana-tropous, bitegmic and tenui-nucellate	"	Linear, chalazal functional	"	Hooked	3–8 antipodals
Eragrostis ciliaris (Bhanwra, 1986; 1988)	Campylo-tropous, bitegmic and tenui-nucellate	"	Linear or T-shaped, chalazal functional	"	Pear-shaped	3 antipodals
E. cilianensis (Swamy, 1944; Kulkarni and Dnyansager, 1984)	Amphi-tropous, bitegmic and tenui-nucellate	Single-celled occasion-ally 2-celled	Linear tetrad, chalazal functional	"	Beaked	9–11 antipodals

Species investigated and Author	Ovule	Archesporium	Megaspore tetrad	Type of embro sac development	Synergids	Antipodals
E. coarctata (Chandra, 1963b; Venkateswarlu and Devi, 1964; Chandra, 1976)	Hemianatropous, bitegmic and tenuinucellate	Single celled	Linear or T-shaped, chalazal functional	"	"	5 antipodals
E. curvula (Rabau *et al.*, 1986)	Bitegmic, tenuinucellate	Single-celled	Linear, chalazal functional	Polygonum type	—	7 antipodals
Eragrostis diarhena (Venkateswarlu & Devi, 1964)	Anatropous, bitegmic and tenuinucellate	"	"	Polygonum type	—	3–6 antipodals
E. minor (Bhanwra, 1988)	"	"	"	"	Hooked	8-antipodals
E. namaquensis (Ghaisas & Patil, 1990)	Campylotropous bitegmic and crassinucellate	—	—	"	—	—
E. nigra (Bhanwra, 1986a; 1988)	Campylotropous, bitegmic and tenuinucellate	Single-celled	Linear or T-shaped, chalazal functional	"	Pear-shaped	3 antipodals
E. pilosa (Chandra, 1976)	Hemianatropous, bitegmic and tenuinucellate	"	Linear or T-shaped	Polygonum type	Beaked	5 antipodals
E. plumosa (Venkateswarlu & Devi, 1964)	Anatropous, bitegmic and tenuinucellate	"	Linear, chalazal functional	"	—	3–6 antipodals
E. poaeoides (Chandra, 1963b; 1976)	Anatropous, bitegmic and tenuinucellate	Single-celled	T-shaped or Linear, chalazal functional	"	Beaked	5 antipodals

(*contd.*)

(*Table II contd.*)

Species investigated and Author	Ovule	Arche-sporium	Megaspore tetrad	Type of embro sac develop-ment	Synergids	Antipodals
E. superba (Streetman, 1963; Shanthamma and Narayan, 1976–77)	—	—	—	Polygonum type	—	3 anti-podals, degenerate
E. tef (Mengesha and Guard, 1966; Longly *et al.*, 1985)	Hemiana-tropous, bitegmic and tenui-nucellate	Single-celled	Linear tetrad, chalazal functional	"	—	11–19 antipodals
E. tenella (Venkateswarlu and Devi, 1964; Bhanwra, 1984; 1988)	Campylo-tropous, bitegmic and tenui-nucellate	"	Linear or T-shaped	"	Pear-shaped	8 anti-podals
E. tremula (Bhanwra, 1986a; 1988)	"	"	"	"	"	6–9 anti-podals
E. unioloides (Untawale *et al.* 1969; Chandra, 1963; 1976)	Hemiana-tropous, bitegmic and tenui-nucellate	Single-celled	Linear or T-shaped, chalazal functional	Polygonum type	Hooked	5 anti-podals
E. viscosa (Venkateswarlu and Devi, 1964; Febulaus and Pullaiah, 1990)	Hemiana-tropous, bitegmic and tenui-nucellate	"	"	"	Pear-shaped	3–6 anti-podals
Eremopogon foveolatus (Satyamurty, 1983; Seshavatharam and Bhaskara Rao, 1984)	Anatropous bitegmic and pesudocrassi-nucellate	"	Linear, chalazal functional	Sexual Polygonum type somatic 4-nucleate aposporus type	—	6-8 antipodals
Eriochloa procera (Venkateswarlu and Devi, 1964)	"	"	"	Sexual Polygonum type Apomictic Aposporous type	—	3 antipodals

Species investigated and Author	Ovule	Archesporium	Megaspore tetrad	Type of embro sac development	Synergids	Antipodals
Erythranthera australis (Verboom *et al.*, 1994)	Hemianatropous, bitegmic and tenuinucellate	—	—	Polygonum type	Synergid haustoria well developed	Antipodals proliferate
E. pumila (Philipson & Connor, 1984)	"	—		—	"	"
Euchlaena mexicana (Cooper, 1937; Koul & Sahi, 1962)	Amphianatropous and bitegmic	Single-celled	—	—	Persistent	30–40 antipodals, 2–4 nucleate persistent
Eustachys glauca (Aulbach Smith & Herr, 1984)	Amphitropous, bitegmic tenuinucellate	Single-celled	Linear T-shaped chalazal functional	Polygonum type	—	6 antipodals
Eustachys. petraea (Aulbach-Smith and Herr, 1984)	"	"	"	"	—	12 antipodals
Festuca microstachys (Maze and Bohm, 1977; Verboom *et al.*, 1994)	Hemianatropous bitegmic and tenuinucellate	Multicelled	T-shaped, chalazal functional	"	—	3 antipodals increase in size
Fingerhuthia africana (Verboom *et al.*, 1994)	"	—	—	"	—	Antipodals proliferate
Glyceria tonglensis (Bhanwra *et al.*, 1982)	Hemianatropous bitegmic and pseduocrassinucellate	Single-celled	Linear triad, chalazal functional	"	Hooked	3 antipodals, cells hypertrophied
Harpochloa falx (Strydom and Spies, 1994)	Amphitropous bitegmic	—	—	"	—	—

(*contd.*)

(*Table II contd.*)

Species investigated and Author	Ovule	Arche-sporium	Megaspore tetrad	Type of embro sac develop-ment	Synergids	Antipodals
Helictotrichon virescens (Bhanwra, 1988)	Hemiana-tropous, bitegmic and tenui-nucellate	"	Linear, chalazal functional "	" "	"	40 antipodals
Hierochloe australis (Weimarck, 1967a)	Anatropous, bitegmic	1–3 celled	Linear, chalazal functional	Sexual Polygonum type, Apomitic-Aposporous	—	"
H. alpina (Weimarck, 1970)	"	"	—	Aposporous 8-nucleate embryo sac	—	"
H. monticola (Weimarck, 1970)	"	1–5-celled	—	"	—	—
H. odorata (Weimarck, 1967a)	Anatropous	1–5-celled	Linear, chalazal functional	Sexual Polygonum Apomictic Aposporous	—	Many antipodals
Hordeum vulgare (Cass and Peteya, 1985)	—	—	Linear tetrad, chalazal functional	Polygonum type	—	—
Imperata cylindrica (Bhanwra, 1982; 1988)	Hemiana-tropous bitegmic and pseudocrassi-nucellate	Single-celled	Linear or chalazal functional	Polygonum type	Flask shaped	8–9 antipodals
Ischaemum rugosum (Sreenivasa Rao, 1997)	Hemiana-tropous bitegmic and pseudocrassi-nucellate	Single-celled	Linear, chalazal functional	Polygonum type	Flask shaped	8-9 antipodals
Iseilema anthephoroides (Venkateswarlu and Devi, 1964)	Anatropous, bitegmic and pseudocrassi-nucellate	"	"	"	—	3–8 antipodals
I. prostratum (Bhanwra *et al.*, 1982;	Hemiana-tropous bitegmic	"	"	"	—	5–8 antipodals cells 6-

Species investigated and Author	Ovule	Arche-sporium	Megaspore tetrad	Type of embro sac develop-ment	Synergids	Antipodals
Bhanwra, 1988)	and pseudo-crassi-nucellate					nucleate
Jansenella griffithinana (2n = 20) (Shobha & Sindhe, 1987)	Campylo-tropous bitegmic and tenui-nucellate	Single celled	Linear chalazal functional	Polygonum type	—	8–10 antipodals
Karrochloa purpurea (Verboom *et al.*, 1994)	Hemiana-tropous, bitegmic and tenui-nucellate	—	—	Polygonum type	Synergid haustoria well developed	Antipodals proliferate
K. schismoides (Verboom *et al.*, 1994)	"	—	—	"	"	"
K. tenella (Verboom *et al.*, 1994)	"	—	—	"	"	"
Lamprothyrsus peruvianus (Verboom *et al.*, 1994)	"	—	—	"	"	"
Leptochloa chinensis (Bhanwra *et al.*, 1981)	Hemiana-tropous bitegemic and tenui-nucellate	Single-celled	"	"	Pear-shaped	5–7 antipodals, uni-nucleate
L. neesi (Venkateswarlu and Devi, 1964)	Anatropous bitegmic and tenui-nucellate	"	"	"	—	More than 10 antipodals
L. panicea (Bhanwra, 1988; Venkateswarlu and Devi, 1964)	"	"	"	"	Hooked	"

(*contd.*)

(*Table II contd.*)

Species investigated and Author	Ovule	Arche-sporium	Megaspore tetrad	Type of embro sac develop-ment	Synergids	Antipodals
Leersia hexandra (Venkateswarlu and Devi, 1964)	Anatropous bitegmic and tenui-nucellate	Single-celled	—	well developed	—	3–4 antipodals
Limnopoa meeboldii (2n = 20) (Shobha & Sindhe, 1992)	Hemiana-tropous, bitegmic and pseudo-rassinu-cellate	Single-celled	Linear triad, chalazal functional	Polygonum type	—	30–40 antipodals, uni-nucleate
Lolium perenne (Bhanwra, 1988)	Hemiana-tropous bitegmic and tenui-nucellate	"	Linear triad, chalazal functional	"	Pear-shaped, hooked	3 antipodals
Melanocenchris jacquemontii (Febulaus and Pullaiah, 1997)	Hemiana-tropous bitegemic and tenui-nucellate	Single-celled	Linear triad, chalazal functional	Polygonum type	Pear-shaped	3 antipodals
Melocalamus compactus (Harigopal and Mohan Ram, 1987)	Hemiana-tropous, bitegmic and Pseuocrassi-nucellate	—	"	"	—	More than 40 antipodals
		—	"	"	—	
Melocanna bambusoides (Stapf, 1904; Petrova, 1965)	Ategmic	Single-celled	—	Polygonum type	Pear-shaped, hooked	40 anti-podals
Merxmuellera arundinacea (Verboom *et al.*, 1994)	Hemiana-tropous, bitegmic and tenui-nucellate	—	—	Apomicitic	Synergid Synergid weak	Antipodals proliferate
M. rufa (Verboom *et al.*, 1994)	"	—	—	Polygonum type	Synergid haustoria well developed	"

Species investigated and Author	Ovule	Archesporium	Megaspore tetrad	Type of embro sac development	Synergids	Antipodals
M. disticha (Verboom *et al.*, 1994)	Hemianatropous, bitegmic and tenuinucellate	—	—	Polygonum type	Synergid haustoria well developed	”
M. dura (Verboom *et al.*, 1994)	”	—	—	”	”	”
M. stricta (Verboom *et al.*, 1994)	”	—	—	”	”	”
Microstegium ciliatum (Bhanwra, 1988)	Hemianatropous, bitegmic and crassinucellate	Single-celled	Linear, chalazal functional	”	—	6–9 antipodals
			”	—		
Ochlandra travancorica (Harigopal and Mohan Ram, 1987)	Hemianatropous bitegmic and pseudocrassinucellate	—	—	Polygonum type	—	More than 40 antipodals
Oplismenus burmanii (Bhanwra and Soni, 1986; Bhanwra, 1988)	”	Single-celled		Linear, chalazal functional	—	5–6 antipodals
O. compositus (Bhanwra and Soni, 1986; Bhanwra, 1988)	”	”	”	”	—	”
Oropetium thomaeum (Diwanji and Diwanji, 1981)	Campylotropous bitegmic crassinucellate	—	—	Polygonum type	—	6–8 uninucleate cells
O. villosulum (Diwanji and Diwanji, 1981)	”	—	—	”	—	”

(*contd.*)

(*Table II contd.*)

Species investigated and Author	Ovule	Archesporium	Megaspore tetrad	Type of embro sac development	Synergids	Antipodals
Oryza alta (Sreenivasa Rao, 1997)	Hemianatropous bitegmic and tenuinucellate	Single-celled	Linear, chalazal functional	Polygonum type	Hooked	3-antipodals
Oryza brachyantha (Sreenivasa Rao, 1997)	"	"	"	"	"	"
Oryza coarctata (Paul and Datta, 1959; 1953)	Campylotropous bitegmic and crassinucellate	Single-celled	"	"	—	3 uninucleate antipodals
O. latifolia (Venkateswarlu and Devi, 1964)	Anatropous bitegmic and tenuinucellate	"	"	"	Hooked	"
Oryza longistaminata (Sreenivasa Rao, 1997)	"	2-celled	T-shaped chalazal functional	"	"	"
Oryza malampuzhaensis (Sreenivasa Rao, 1997)	"	Single celled	Linear chalazal functional	"	"	"
Oryza minuta (Sreenivasa Rao, 1997)	"	2-celled	"	"	"	3-antipodals, cells 3–5, nucleate
Oryza officinalis (Sreenivasa Rao, 1997)	"	Single celled	"	"	"	3-antipodals
Oryza rhizomatis (Sreenivasa Rao, 1997)	"	Single celled	"	"	"	3–5 antipodals
Oryza ridleyi (Sreenivasa Rao, 1997)	"	Single celled	"	"	"	3–5 antipodals

Species investigated and Author	Ovule	Arche-sporium	Megaspore tetrad	Type of embro sac develop-ment	Synergids	Antipodals
Oryza rufipogon (Sreenivasa Rao, 1997)	Anatropous bitegmic and tenui-nucellate	Single celled	Linear chalazal functional	Polygonum type	Hooked	3 antipodals
O. sativa (Kuwada, 1910; Terada, 1928; Santos, 1933; Juliano and Aldama, 1937; Raju, 1980; Bhanwra, 1988)	Hemiana-tropous, bitegmic and tenui-nucellate	Single-celled	Linear, chalazal functional	Polygonum type	Hooked	3–4 anti-podals
Oryzopsis asperifolia (Kam and Maze, 1974)	"	"	"	"	Hooked filiform apparatus present	More than 3 antipodals
O. hendersoni (Mehlenbacher, 1970)	"	Multi-cellular	T-shaped	"	—	—
O. hymenoides (Kam and Maze, 1974)	"	Single-celled	Commonly linear, rarely T-shaped, chalazal functional	"	"	"
O. kingii (Kam and Maze, 1974)	"	"	T-shaped, chalazal functional	"	"	More than 3 anti-podals
O. micrantha (Kam and Maze, 1974)	"	"	Linear tetrad, chalazal functional	"	"	"
O. miliacea (Maze *et al.*, 1970; 1971)	Hemiana-tropous, bitegmic and pseudocrassi-nucellate	Single-celled, ccasion-ally two-celled	T-shaped or linear tetrad, chalazal functional	"	—	"
Panicum maximum (Warmke, 1954)	"	—	Linear tetrad or triad, chalazal functional	Sexual Polygonum	—	—

(*contd.*)

(*Table II contd.*)

Species investigated and Author	Ovule	Arche-sporium	Megaspore tetrad	Type of embro sac develop-ment	Synergids	Antipodals
P. miliaceum (Narayanaswami, 1955a)	Hemiana-tropous, bitegmic and crassi-nucellate	Single-celled	Row of three cells or T-shaped	Polygonum type	Pyriform, hooked	6 anti-podals; each 1-4-nucleate
P. milliare (Narayanaswami, 1955a)	"	"	Linear tetrad or row of three cells	"	Pear-shaped	3 anti-podals
P. notatum (Febulaus, Pullaiah, 1992a)	Hemiana-tropous bitegmic and pseudocrassi-nucellate	"	Linear tetrad all mega-spores degenerate	Embryo sac aposporous	Hooked	3 anti-podals;
P. repens (Febulaus and Pullaiah, 1992a)	"	"	Linear or T-shaped chalazal functional	"	Pear-shaped	"
Paspalidium flavidum (Bhanwra *et al.*, 1982; Bhanwra, 1988)	Hemiana-tropous, bitegmic and pseudocrassi-nucellate	Single-celled	Linear or chalazal functional	Polygonum type	—	5–8 anti podals;
Paspalum alcalinum (2n = 20) (Burson, 1997)	—	"	"	Polygonum type	—	—
Paspalum arechavaleate (Burson & Bennet, 1971a)	—	—	"	Polygonum type	—	—
Paspalum boscianum (Quarin & Hanna, 1980)	—	—	Linear, Chalazal functional	Polygonum type	—	Antipodals prolifera-ted
Paspalum chacoence (2n = 20) (Burson, 1985)	—	—	"	"	—	—

Species investigated and Author	Ovule	Arche-sporium	Megaspore tetrad	Type of embro sac develop-ment	Synergids	Antipodals
Paspalum commersonii (Chao, 1974)	—	—	—	Diplosporous in Hexaploid and in induced dodecaploid, Polygonum type in Dodecaploid Biotype.	—	3 anti-podals
Paspalum conspersum (2n = 4X = 40) (Burson & Bennett, 1976; Quarin & Hanna, 1980)	—	—	Linear, chalazal functional	Polygonum type	—	Antipodals proliferate
Paspalum dasyplerum (Quarin & Caponio, 1995)	—	—	—	Polygonum type	—	Antipodals proliferate
Paspalum densum (2n=20) (Caponio & Quarin, 1993)	—	—	Linear, chalazal functional	Polygonum type	—	Antipodals proliferate
Paspalum dilatatum (2n = 40) (Bashaw & Holt, 1988)	—	—	"	"	—	—
P. distichum (Bhanwra, 1988)	Hemiana-tropous, bitegmic and pesudocrassi-nucellate	—	Linear, chalazal functional	Sexual embryo sac degenerates, unreduced 8-nucleate embryo sac	—	5–6 anti-podals
Paspalum guaraniticum (2n = 40) (Burson & Bennet, 1970b)	—	—	"	Polygonum type		
P. guenorum (Pritchard, 1970)	—	—	Apomietic			
Paspalum hexstachyum	—	—	—	Polygonum type	—	—

(*contd.*)

(*Table II contd.*)

Species investigated and Author	Ovule	Arche-sporium	Megaspore tetrad	Type of embro sac develop-ment	Synergids	Antipodals
(2n = 60) (Pitman *et al.*, 1987; Quarin & Hanna, 1980)						
Paspalum intermedium (Burson and Bennet, 1970b)	—	Single-celled	Linear, chalazal functional	Polygonum type	—	—
Paspalum jurgensii (Burson and Bennet, 1971b)	—	Single-celled	"	"	—	—
Paspalum laxum (Quarin *et al.*, 1982)	—	—	—	Polygonum type	—	—
P. longifolium (Chao, 1974)	—	—	—	Diplosporous in tetraploid and Polygonum type in induced octoploid	—	Many anti-podals in octoploid type cells multi-nucleate
Paspalum malacophyllum (2n = 40) (Burson & Hussey, 1998)	—	—	Linear chalazal functional	Polygonum type	—	—
Paspalum modestum (2n = 20) (Burson, 1997; Quarin & Hanna, 1980)	—	Single-celled	"	"	—	Antipodals proliferal
Paspalum monostachyums (2n = 20) (Burson, 1997)	—	"	"	"	—	"
Paspalum paniculatum (Burson & Bennett, 1971b)	—	Single-celled	"	"	—	—

Species investigated and Author	Ovule	Arche-sporium	Megaspore tetrad	Type of embro sac develop-ment	Synergids	Antipodals
Paspalum platyphyllum (Bashaw *et al.*, 1970)	—	—	—	Polygonum type	—	—
P. plicatulum (Pritchard, 1970)	—	—	—	Unreduced 8-nucleate embryo sac	—	—
Paspalum pubiflorum (Bashaw *et al.* 1970)	—	—	—	Polygonum type	—	—
Paspalum pumilum (Burson & Bennet, 1971a)	—	—	Linear, chalazal functional	Polygonum type	—	—
Paspalum regnelli (Norrmann, 1981)	—	—	—	Polygonum type	—	—
Paspalum repens (2n = 20) (Burson, 1997)	—	Single-celled	Linear, chalazal functional	Polygonum type	—	Antipodals proliferal
P. scrobiculatum (Narayanaswami, 1954)	Anatropous bitegmic and crassi-nucellate	Single-celled	Linear tetrad	Polygonum type	Pear-shaped	As many as 15 cells each with 1-4 nuclei
P. secans (Pritchard, 1970)	—	—	—	Unreduced 8-nucleate embryo sac	—	—
Paspalum urvillei (Bashaw *et al.*, 1970)	—	—	Linear chalazal functional	Polygonum type	—	—
Paspalum virgatum (Burson & Quarin, 1982)	—	—	"	"	—	—
Pennisetum ciliare (Snyder *et al.*, 1955)	—	—	Linear tetrad or triad, chalazal functional	Sexual Polygonum type	—	—

(*contd.*)

(*Table II contd.*)

Species investigated and Author	Ovule	Arche-sporium	Megaspore tetrad	Type of embro sac develop-ment	Synergids	Antipodals
Pennisetum macrosatchyum (Shanthamma, 1979)	—	Single-celled	Linear tetrad chalazal functional	"	Hooked	6–12 anti-podals uni-nucleate or multi-nucleate
P. orientale (Choda *et al.*, 1981)	Hemiana-tropous bitegmic and pseudocrassi-nucellate	"	Linear, all degenerate	Aposporus, 4-nucleate	—	4–6 anti-podals 12–24 cells
Pennisetum pedicellatum (2n = 34) (Shobha, 1988)	Hemiana-tropous, bitegmic and tenui-nucellate	—	—	Aposporous	—	—
Pennisetum purpureum (Saran & Misra, 1986)	—	—	Linear chalazal functional	Polygonum type	—	—
P. typhoideum (Rangaswamy, 1935; Krishnaswamy and Ayyangar, 1941; Khosla, 1946; Narayanaswami, 1953; Raju, 1980)	Campylo-tropous, bitegmic and crassi-nucellate	Single-celled	Linear tetrad, chalazal functional	Polygonum type	Pear-shaped	5–10 each with 3–8-nuclei, persistent
Pentaschistis airoides (Spies *et al.*, 1999)	Anatropous bitegmic	—	—	Polygonum type	—	—
Pentaschistis ampla (Verboom *et al.*, 1994)	Hemiana-tropous, bitegmic and tenui-nucellate	—	—	Polygonum type	Synergid haustoria weak	Antipodals proliferate
P. argentea (Verboom *et al.*, 1994)	"	—	—	"	—	"
P. aristifolia (Verboom *et al.*, 1994)	"	—	—	"	—	"

Species investigated and Author	Ovule	Arche-sporium	Megaspore tetrad	Type of embro sac develop-ment	Synergids	Antipodals
P. aristoides (Verboom *et al.*, 1994)	Hemiana-tropous, bitegmic and tenui-nucellate	—	—	Polygonum type	Synergid haustoria weak	Antipodals proliferate
Pentaschistis calcicola (Spies *et al.*, 1999)	Anatropous bitegmic	—	—	Polygonum type	—	—
P. chippendalliae (Verboom *et al.*, 1994)	"	—	—	"	Synergid haustoria well developed	"
P. curvifolia (Spies *et al.*, 1999)	Anatropous	—	—	Polygonum type	—	—
P. eriostoma (Verboom *et al.*, 1994)	"	—	—	"	"	"
P. pallescens (Verboom *et al.*, 1994)	"	—	—	"	Synergid haustoria weak	
Pentaschistis pallida (Spies, *et al.*, 1999)	Anatropus bitegmic	—	—	Polygonum type	—	—
P. patula (Verboom *et al.*, 1994)	"	—	—	"	Synergid haustoria well developed	
P. pungens (Verboom *et al.*, 1994)	"	—	—	"	—	"
Pentaschistis setifolia (Spies *et al.*, 1999)	Anatropous bitegmic	—	—	Polygonum type	—	—
P. tomentella (Verboom *et al.*, 1994); (Spies, *et al.*, 1999)	Hemiana-tropous, bitegmic and tenui-nucellate	—	—	Polygonum type	—	Antipodals proliferate

(contd.)

(*Table II contd.*)

Species investigated and Author	Ovule	Arche-sporium	Megaspore tetrad	Type of embro sac develop-ment	Synergids	Antipodals
Pentaschistis tortuosa (Spies *et al.*, 1999)	Anatropous	—	—	Polygonum type	—	—
P. velutina (Verboom *et al.*, 1994)	"	—	—	"	Synergid haustoria well developed	"
Pentaschistis viscidula (Spies *et al.*, 1999)	Anatropous bitegmic	—	—	Polygonum type	—	—
Pentanesis macrocalycina (Verboom *et al.*, 1994)	"	—	—	"	Synergid haustoria well developed	
Perotis hordieformis (Venkateswarlu and Devi, 1964; Bhanwra, 1988)	Hemiana-tropous, bitegmic and tenui-nucellate	Single-celled	Linear, chalazal functional	Polygonum type	Hooked	3 anti-podals
P. indica (Febulaus and Pullaiah, 1993a)	"	"	T-shaped, chalazal functional	"	"	"
Phalaris minor (Bhanwra, 1988)	"	"	Linear, chalazal functional	"	"	"
Phragmites communis (Satyamurty and Seshavatharam, 1984)	Campylo-tropous, bitegmic and tenui-nucellate	"	"	"	"	"
P. karka (Bhanwra, 1988)	"	"	"	"	Hooked	6 anti-podals
Poa alpina (Hakanson, 1943)	—	—	Linear, chalazal functional	Sexual Polygonum type and apomictic aposporous type	—	—

Species investigated and Author	Ovule	Arche-sporium	Megaspore tetrad	Type of embro sac develop-ment	Synergids	Antipodals
P. annua (Chandra, 1963b; Bhanwra, *et al.*, 1985; Bhanwra, 1988)	Nearly othrotropous, bitegmic and tenui-nucellate	Single-celled	Linear, T-shaped, chalazal functional	Polygonum type	Pear-shaped	5–7 anti-podals
P. arctica (Engelbert, 1941)	—	—	—	Both sexual Polygonum and aposporous type	—	—
P. compressa (Anderson, 1927)	Campylo-tropous, and bitegmic	Single-celled, occasion-ally 2-celled	Linear, chalazal functional, occassinally micropylar functional or all degene rate	Sexual embryo sacs absent, aposporous embryo sacs are formed	—	Antipodals increase in numer
Polypogon fugax (Bhanwra *et al.*, 1981; Bhanwra, 1988)	Hemiana-tropous, bitegmic and tenui-nucellate	Single-celled, occasio-nally 2-celled	Linear, or T-shaped, chalazal functional	Polygonum type	Pear shaped	3 anti-podals
P. monspe-liensis (Bhanwra *et al.*, 1981; Bhanwra, 1988)	"	"	"	"	"	"
Pseudo-pentameris macrantha (Verboom *et al.*, 1994)	"	—	—	"	Synergid haustoria well developed	
Pseudosasa japonica (Verboom *et al.*, 1994)	"	—	—	"	—	"
Pseudo-stachyum polymorphum (Harigopal and Mohan Ram, 1987)	Hemiana-tropous, bitegmic and pseudocrassi-nucellate	—	—	Polygonum type	—	More than 40 anti-podals

(*contd.*)

(*Table II contd.*)

Species investigated and Author	Ovule	Arche-sporium	Megaspore tetrad	Type of embro sac develop-ment	Synergids	Antipodals
Pyrrhanthera exigua (Philipson & Connor, 1984)	"	—	—	—	—	"
Rostraria phleoides (Bhanwra, 1988)	"	"	"	"	"	"
Rottboellia exaltata (Sreenivasa Rao, 1997)	Hemiana-tropous, bitegmic and pseudo-crassi-nucellate	Single celled	Linear, chalazal functional	Polygonum type	Flask shaped	3 antipodals
Rytidosperma clelandii (Verboom *et al.*, 1994)	"	—	—	"	Synergid haustoria well developed	"
R. erianthum (Verboom *et al.*, 1994)	"	—	—	"	"	"
R. setifolium (Philipson & Connor, 1984)	"	—	—	"	"	"
R. vickeryi (Verboom *et al.*, 1994)	"	—	—		"	"
Saccharum benghalense (Bhanwra *et al.*, 1982; Bhanwra, 1988)	Hemiana-tropous bitegmic and pseudo crassinucellate	Single-celled, rarely two-celled	Linear, chalazal functional	Polygonum type	Flask shaped	5–8 anti-podals 6-nucleate
S. officinarum (Narayanaswami, 1940)	—	—	—	Unreduced 8-nucleate embryo sac	—	—
S. spontaneum (Narayanaswami, 1940)	—	—	—	"	—	—

Species investigated and Author	Ovule	Archesporium	Megaspore tetrad	Type of embro sac development	Synergids	Antipodals
Sasa paniculata (Yamaura, 1933)	—	—	—	—	—	399 antipodals
Schmidtia pappophoroides (Verboom *et al.*, 1994)	Hemianatropous, bitegmic and pseudocrassinucellate	Single-celled, rarely two-celled	"	"	—	5-8 antipodals 6-nucleate
Secale cereale (Xi and Demason, 1984)	Bitegmic, tenuinucellate	Single-celled	T-shaped or linear, chalazal functional	Polygonum type	Pear-shaped	3 antipodals
Setaria glauca (Bhanwra *et al.*, 1982; Bhanwra, 1988)	Hemianatropous, bitegmic and pseudocrassinucellate	"	"	"	—	3 amtipodals, each 1-4-nucleate
S. intermedia (Bhanwra *et al.*, 1982; Bhanwra, 1988)	"	"	"	"	—	About 23-27 antipodals
Setaria pumila (Sreenivasa Rao, 1997)	"	"	"	"	Peer shaped	10-12 antipodals
S. italica (Khosla, 1946)	Hemianatropous, bitegmic and pseudocrassi-	Single celled	Linear chalazal functional	Polygonum type	Peer-shaped with short hooks and beaks	10-12 antipodals, each 2-3-nucleate
Setaria pumila (Sreenivasa Rao, 1997)	Hemianatropous, bitegmic and pseudocrassi-	Single celled	Linear chalazal functional	Polygonum type	Pear shaped	10–12 antipodals
S. verticillata (Bhanwra *et al.*, 1982; Bhanwra, 1988)	"	"	Linear or T-shaped, chalazal functional	"	—	5–8 antipodals

(contd.)

(*Table II contd.*)

Species investigated and Author	Ovule	Arche-sporium	Megaspore tetrad	Type of embro sac develop-ment	Synergids	Antipodals
Sieglingia decumbens (Verboom *et al.*, 1994)	Hemiana-tropous, bitegmic and pseudocrassi nucellate	—	—	—	Synergid haustoria well developed	Antipo-dals proliferate
Sorghum arundinaceum (Venkateswarlu and Devi, 1964)	Anantropous bitegemic and pseudo-crassi-nucellate	Single-celled	Linear or T-shaped, chalazal functional	Polygonum type	—	More than 10 anti-podals
S. vulgare (Artschwager and McGuire, 1949)	Hemiana-tropous bitegmic and pseudocrassi-nucellate	"	—	Polygonum type	—	—
Spinifex littoreus (Jayalakshmi and Lakshmanan, 1982)	Anatropous, unitegmic and crassi-nucellate	"	Linear tetrad, chalazal functional	Polygonum type	—	10–15 cells
Sporobolus coromandeli-anus (Seshavatharam and Satyamurty, 1982a; Satyamurty, 1983)	Campylo-tropous, bitegmic and tenui-nucellate	—	Linear and T-shaped	"	—	—
S. diander (Bhanwra *et al.*, 1981; Bhanwra, 1988)	Hemiana-tropous, bitegmic and tenui-nucellate	Single-celled	Linear tetrad, chalazal functional	Polygonum type	Pear-shaped	3-6 uni-nucleate antipodals
S. fertilis (Bhanwra, 1988)	"	"	"	"	"	3-anti-podals
Sporobolus indicus (Sreenivasa Rao, 1997)	"	"	Linear chalazal functional	"	"	"

Species investigated and Author	Ovule	Arche-sporium	Megaspore tetrad	Type of embro sac develop-ment	Synergids	Antipodals
Sporobolus spicatus (Sreenivasa Rao, 1997)	"	"	"	"	"	"
S. tremulus (Seshavatharam and Bhaskar Rao, 1983)	Anatropous, bitegmic and tenui-nucellate	"	—	Polygonum type	—	—
Stipa elmeri (Maze and Bohm, 1973; Maze and Lin, 1975)	Hemiana-tropous bitegmic and pseudocrassi-nucellate	Single-celled, occasion-ally 2-celled	T-shaped, chalazal functional	"	—	Antipodals proliferate up to 140 cells
S. lemmonii (Maze *et al.*, 1972)	"	Single-celled	Linear, chalazal functional	"	—	102 anti-podals at 2-celled proembryo-satage
S. leucotricha (Brown, 1949)	Bitegmic and crassi nucellate	—	"	"	"	Antipodals increase in number, multi-nucleate
S. tortilis (Maze *et al.*, 1970)	Hemiana-tropous, bitegmic and pseudocrassi-nucellate	Single-celled	Linear	"	—	Antipodals proliferate
Themeda cymbaria (Muniyamma, 1973)	—	Single-celled occasion-ally 2-celled	—	—	—	—
Themeda triandra (Merwe, 1957)	Campylo-tropous, bitegmic and crassi-nucellate	—	Linear tetrad, chalazal functional	Diploids sexual Polygonum type	—	Increase in number of cells and nuclei

(*contd.*)

(*Table II contd.*)

Species investigated and Author	Ovule	Archesporium	Megaspore tetrad	Type of embro sac development	Synergids	Antipodals
Thyrsostachys oliveri (Bhanwra, 1988)	Hemianatropous, bitegmic and pseudocrassinucellate	Single-celled, rarely 2-celled	Linear or T-shaped, chalazal functional	Polygonum type	Per shaped and hooked	12 antipodals
Trachys muricata (Venkateswarlu and Devi, 1964)	Anatropous bitegmic and pseudocrassinucellate	Single-celled	Linear, chalazal functional	"	—	3–8 antipodals
Tragus bertorianus (Strydom and Spies, 1994)	—	—	—	Polygonum type	—	—
Tragus biflorus (Chandra 1963b; 1975; Bhanwra *et al.*, 1981)	Hemianatropous, bitegmic and tenuinucellate	"	T-shaped or linear, chalazal functional	"	Hooked	3 uninucleate antipodals
Tribolium acutiflorum (Visser & Spies, 1994)	Anatropous bitegmic and tenuinucellate	—	—	Polygonum type	—	—
Tribolium brachystachum (Visser & Spies, 1994)	Anatropous bitegmic and tenuinucellate	—	—	Polygonum type	—	—
Tribolium ciliare (Visser & Spies, 1994)	Anatropous bitegmic and tenuinucellate	—	—	Polygonum type	—	—
Tribolium echinatum (Visser and Spies, 1994)	Anatropous, tenuinucellate	—	—	Polygonum type	—	—
Tribolium glomeratum (Visser & Spies, 1994)	Anatropous bitegmic and tenuinucellate	—	—	Polygonum type	—	—

Species investigated and Author	Ovule	Arche-sporium	Megaspore tetrad	Type of embro sac develop-ment	Synergids	Antipodals
Tribolium hispidum (Visser & Spies, 1994)	Anatropous bitegmic and tenui-nucellate	—	—	Polygonum type	—	—
Tribolium obliterum (Verboom *et al*, 1994)	Hemiana-tropous, bitegmic and tenui-nucellate	—	—	Polygonum type	Synergid haustoria well developed	Antipodals proliferate
Tribolium uniolae (Visser and Spies, 1994)	"	—	—	"	—	—
Tribolium urticulosum (Visser and Spies, 1994)	"	—	—	"	—	—
Tripogon filiformis (Bhanwra *et al*., 1981; Bhanwra, 1988)	Hemiana-tropous, bitegmic and tenui-nucellate	Single-celled	Linear or T-shaped, chalazal functional	Polygonum type	Pear-shaped	10-antipodals
Tripsacum dactyloides (Farquharson, 1955)	—	—	—	Unreduced 8-nucleate embryo sac	—	—
Triticum aestivum (Bhatnagar and Chandra, 1975; Aziz, 1972; Xi Cüi, 1983; Raju, 1980)	Anatropous, bitegmic and tenui-nucellate	2–3 celled	Linear, chalazal functional	Polygonum type	Hooked persists till 4-6-celled embryo stage	15–18 antipodals
Triticum dicoccum (Raju, 1980)	Campylo-tropous, bitegmic and crassi-nucellate	1–3 cells	Linear, chalazal functional	Polygonum type	Hooked	20 antipodals

(*contd.*)

(Table II contd.)

Species investigated and Author	Ovule	Archesporium	Megaspore tetrad	Type of embro sac development	Synergids	Antipodals
T. durum (Raju, 1980)	"	1–2 cells	Linear or T-shaped, chalazal functional	"	"	"
T. vulgare (Percival, 1921)	Bitegemic	—	—	—	—	Polyploid, occasionally up to 28
Urochlaena pusilla (Verboom *et al.*, 1994)	Hemianatropous, bitegmic and tenuinucellate	—	—	Polygonum type	Synergid haustoria well developed	
Urochloa oligotricha (Verboom *et al.*, 1994)	Hemianatropous, bitegmic and tenuinucellate	—	—	—	—	Antipodals proliferate
Urochloa panicoides (Basavaiah and Murty, 1988)	Hemianatropous, bitegmic and pseudocrassinucellate	Single-celled	Linear or rarely T-shaped, chalazal functional	Polygonum type	—	6–12 antipodals
Vetiveria zizanioides (Bhanwra, *et al.*, 1982; Bhanwra, 1988)	Hemianatropous bitegmic and pseudocrassinucellate	Single-celled	Linear or rarely T-shaped, chalazal functional	Polygonum type	—	6–12 antipodals
Zea mays (Miller, 1919; Randolph, 1936; Cooper, 1937; Raju, 1980)	Amphianatropous, bitegmic	"	Linear	"	Persistent	30–40 antipodals, 2–4-nucleate persistent
Zingeria trichopoda (Shehata, 1995)	Campylotropous, bitegemic and tenuinucellate	Single-celled	Linear chalazal functional	Polygonum type	Persistent	6 cells each binucleate

CHAPTER 5

FERTILIZATON

The process of fusion of male gamete with a female gamete is described as fertilization. The pollen grain on germination gives rise to pollen tube. The pollen tube grows through the style and enters the ovule through micropyle. It continues to grow into the embryo sac. On entering the embryo sac the vegetative nucleus disintegrates and the two sperms are discharged in the vicinity of the egg. One of the sperm nuclei fuses with the egg to form zygote while the other unites with the polar nuclei to form primary endosperm nucleus.

The Stigma

In grasses the stigma is described as dry stigma. In *Zea*, *Sorghum* and *Hordeum* these dry stigmas have dispersed receptive cells on multiseriate branches. The dry stigma has a cuticle covered with a pellicle. The pellicle is an extracellular product originating from the stigmatic cells of a hydrated protein film (Van Went and Willemse, 1984). The stigmatic papillae arise from the epidermal cells of the stigma. In *Secale* the papillate wall shows six layers - a peillicle, a mucilage layer containing mucopolysaccharides, a cuticle and a thick layer with microfibrillar content containing pectine. The last layer adjoins a layer with one to four linked glucans. Vesicles with proteins are present in the wall (J. Helsop-Harrison and Y. Helsop-Harrison, 1980). The nature of papillate walls show great diversity

Pollen Germination and Pollen Tube Growth

The time taken by the pollen grain to germinate on the stigma is quite variable. Very little information is available on this aspect. For example in *Zea mays* (Randolph, 1936) and *Hordeum distichon* (Pope, 1937) it takes only 5 minutes. In *Saccharum officinarum* (Artschwager *et al.*, 1929) and *Sorghum vulgare* (Artschwager and Mcguire, 1949) germination of pollen takes place immediately after reaching the stigma surface.

The stigmatic exudate provides a suitable medium for pollen germination. Lipids, one of the chief constituents in *Zea mays* (Martin, 1970) serve the function of "liquid cuticle" and protect the pollen from excessive transpiration and dessiccation.

The pollen on germination gives rise to pollen tube. The pollen tube first grows along the surface of the hair like cells and then penetrates its wall. In barley the pollen tube utilizes its own reserve substances for penetration. Following penetration of the cell wall, a thickened and looser texture of the wall is visible after the excretion of enzymes by the pollen tube tip (Cass and Peteya, 1979). The style in Poaceae is of the soild type. In *Paspalum* (Chao, 1971) the pollen tubes grow through the intercellular substance of the transmitting tissue of solid style. The pollen tube grows down from the style into the intercellular space between the ovary wall and the outer integument. The distance a pollen tube has to travel in order to reach the egg depends on the length of the style which is quite variable in different species. In *Zea mays* it has to grow as much as 450 mm.

Entry of Pollen Tube into the Ovule

When the pollen tube reaches the micropylar end of the ovule, it touches the PAS-substance by which it grows into the micropyle and intercellular spaces between rows of nucellar cells below the embryo sac.

The growth of pollen tube towards the embryo sac is associated with the chemotrophic substances present in the ovule. Synergids are believed to be the origin of chemotrophic substances involved in directing the growth of pollen tube towards the embryo sac. According to Diboll and Larson (1966) the synergids of *Zea mays* contain an abundant number of mitochondria and dictyosomes, a well developed endoplasmic reticulum with partly parallel arranged cisternae and numerous ribosomes which indicate a high metabolic activity. These synergids are supposed to produce the chemotrophic substances into the filiform apparatus. From filiform apparatus the chemotropic substances enter the micropyle (Van Went and Willemse, 1984).

Chao (1971) has carried out a detailed cytological investigation on *Paspalum orbiculare* and reported that the distal part of the integument, by dissolution of its cells *in situ*, secretes mucilaginous substance into the micropyle which provides a way of least resistance for the pollen tube and guides it towards its ultimate destination. This mucilaginous substance is largely a water soluble carbohydrate and it aids the pollen tube growth both mechanically and chemotrophically.

The time between pollination and fertilization is different in different species. It depends on the rate of the growth of pollen tube and the distance it travels through the stylar tissue. In *Oryza sativa* (Juliano and Aldama, 1937) fertilization takes place in about 12 to 14 hours after pollination. In *Bromus innermis* (Nielsen, 1947) it takes in about 15 to 18 hours. In *Hordeum distichon-palmella*

(Pope, 1937) the pollen tube arrives inside the embryo sac in less than an hour after pollination. In *Triticum* (Bhatnagar and Chandra, 1978) fertilization occurs 10-16 hours after pollination. In *Dicanthium armatum* fertilization occurs between one to two hours after pollination (Reddy and D'Cruz, 1969c).

Entry of Pollen Tube into the Embryo Sac

The mode of pollen tube entry into the embryo sac differs in different species. It penetrates the embryo sac wall and enters the degenerating synergid through the upper part of the filiform apparatus in *Paspalum conjugatum* (Chao, 1980). In *Stipa elmeri* (Maze and Lin, 1975), *Hordeum vulgare* (Cass and Jensen, 1970), *Stipa hendersonii* (Mehlenbacher, 1970) and in *Paspalum* (Chao, 1971) the pollen tube enters at the base of the filiform apparatus of one of the synergids. In *Hordeum distichum* (Luxova, 1967) pollen tube entered embryo sac between the wall of the embryo sac and one or both the synergids. In *Zea* Cooper (1937) reported that the pollen tube grows between the two synergids. According to Cho (1956) in *Oryza sativa* the pollen tube enters the embryo sac between one synergid and egg cell. Bhatnagar and Chandra (1975) who studied *Triticum aestivum* also reported that the pollen tube enters the embryo sac between one synergid and egg cell. However according to Bhanwra *et al.* (1981) the pollen appears to enter directly into one of the synergids and discharges its contents. Cass and Jensen (1970) reported that in *Hordeum* the pollen tube penetrates the persistent synergid and discharges its contents into this penetrated synergid. Kam and Maze (1974) also reported that in *Oryzopsis asperifolia* the pollen tube seems to enter the embryo sac via persistent synergid, from which dense, chromatin like bodies appear to be extruded.

The degeneration of synergids is associated with the entry of pollen tube. Shortly before or after the entry of the pollen tube into the embryo sac, one of the synergids show signs of degeneration. This synergid is known as degenerated synergid where as the other is known as persistent synergid. The time of synergid degeneration varies. The synergid begins to degenerate before the pollen tube entry as in *Hordeum vulgare* (Cass and Jensen, 1970), *Zea mays* (Diboll, 1968), *Sorghum* (Vazart, 1955) and *Panicum miliare* (Narayanaswami, 1955) or soon after the pollen tube entry as reported by Maze and Lin (1975). However Cooper (1937) reported that in *Zea* the pollen tube grows between the two synergids without destroying any of them. The degenerated synergid is mostly electron dense (Maze and Lin, 1975) and revealed the presence of bodies of ribonucleic acid and protein associated with the degenerated synergid. The synergid receiving the pollen tube degenerated and that which did not, remained healthy, as in *Zea mays* (Korobova, 1959).

Pollen Tube Discharge

The pollen tube discharges two sperms, the vegetative nucleus and fair amount of tube cytoplasm in the degenerated synergid. The sperms are true cells and are bounded by a plasma membrance. In *Secale* (Karas and Cass, 1976) during the cytokinesis the generative cell is attached to the pollen wall and cell plate formation is clearly visible. The two sperm cells have large nuclei. This nucleus is surrounded by a layer of cytoplasm with mitochondria, dictyosomes, endoplasmic reticulum and parallel array of microtubles.

The sperm cells change their shape in response to conditions of squash preparations (Cass, 1973). In *Hordeum* Cass (1973) reported that the microtubules are responsible for the preservation of the cell shape. The sperm cells are either spherical or spindle in shape. This facilitates them to move into the pollen tube. However Navashin (1969) considered that the movement of sperm cell is due to cytoplasmic streaming. Korobova (1974) conducted *in vitro* studies on the pollen of *Zea* and suggested that the independent movement of vegetative nucleus and the sperm cells are related to a shift of the dynamic centre of the cells. Luxova (1967) noticed that in Barley the contents of the pollen tube are most frequently discharged between the egg cell and polar nuclei.

Post-pollination Changes in the Embryo Sac

The synergid visited by the pollen tube shows signs of degeneration. Its vacuoles disappear and the organells get disorganised (Bhojwani and Bhatnagar, 1974). Darkly stained bodies appear in the cytoplasm of egg and degenerated synergid.

X-bodies

After the pollen tube discharges, in the cytoplasm of the degenerated synergid sometimes darkly stained bodes have been observed by several embryologists. These bodies are termed as X-bodies (Maheshwari, 1950; Fisher and Jensen, 1969; Kapil and Bhatnagar, 1975; Narayanaswami, 1955). Such X-bodies have been reported in barley (Cass and Jensen, 1970), *Stipa elmeri* (Maze and Bohm, 1973), *S. lemmonii* (Maze *et al.*, 1972), *S. tortilis* (Maze *et al.*, 1970), *Echinochloa frumentacea* (Narayanaswami, 1955) and in *Paspalum conjugatum* (Chao, 1980). The X-bodies have been variously interpreted. Some believe that they are the remains of degenerated synergid nucleus or vegetative nucleus. This interpretation is based on their position as well as the presence of DNA (Fisher and Jensen, 1967). Some others are of the opinion that these darkly stained bodies are the cytoplasmic remains of the sperm (Chao, 1980). In *Paspalum conjugatum* Chao (1980) observed four darkly stained bodies. Of these four, he believed the two larger bodies as the nuclei of degenerated synergid and vegetative cell, whereas the two smaller ones represent the discharged cytoplasmic sheaths of the sperm. However the exact nature of these X-bodies is unknown and it needs more detailed study.

Extra bodies such as those observed in the degenerated synergid are also seen on the surface of the egg. In *Echinochloa frumentacea* (Narayanaswami, 1955b) reported that X-bodies arising from the burst tip of the pollen tube form crescent around in the egg cell. Bhanwra *et al.* (1981) also observed such crescent granules both in synergid and egg cell of *Echinochloa colonum* and *E. crusgalli.* Reddy and D'Cruz (1969c) noticed the presence of numerous black granules arranged in the form of crescent between the egg and the endosperm nucleus. Artschwager and Mc Guire (1949) also observed such granules in *Sorghum vulgare*. They believe these bodies as remnants of pollen tube discharge.

Double Fertilization

The pollen tube discharges two sperms. These two sperms fertilize two different cells of the embryo sac. Nawaschin (1898) for the first time reported that the two sperms are involved in fertilization. Of the two sperms one fuses with the egg nucleus and this process is known as syngamy. The other sperm moves towards the central cell and fuses with the fusion product of the polars. This results in the formation of primary endosperm nucleus and the process is known as triple fusion. In aposporic taxa syngamy is absent and the egg develops parthenogenetically. However triple fusion results in the formation of endosperm. In rare cases for example in *Cymbopogon nardus* and *C. martinii* the pollen tubes fails to reach the embryo sac. In this species even in the absence of fertlization endosperm develops autonomously.

According to Gerassimova-Navashina (1960) the type of syngamy of grasses comes under premitotic type where the sperm nucleus fuses immediately on coming in contact with the egg nucleus and the zygote nucleus divides subsequently.

Syngamy and triple fusion occurs simultaneously as in *Polypogon monspeliensis* (Bhanwra *et al.*, 1981), *Paspalum scrobiculatum* (Narayanaswami, 1954), *Dicanthium armatum* (Reddy and D'Cruz, 1969c), *Triticum* (Bhatnagar and Chandra, 1975), *Setaria italica* (Narayanaswami, 1956) etc. or Syngamy precedes triple fusion as in *Sporobolus diander* (Bhanwra *et al.,* 1981), or triple fusion precedes syngamy as in *Eragrostis ciliensis* (Kulkarni and Dnyansagar, 1984). However in several cases syngamy and triple fusion occur simultaneously.

Very little information is available on the mode transfer of sperm cells from the penetrated synergid. Cass and Jensen (1970) reported that in barley only the sperm nuclei have been identified in either the egg cell or the central cell. There is no evidence of presence of male cytoplasm inside the female gametophyte. However Cass (1981) reported the entry of male cytoplasm into the egg cell in barley which is also supported by Carroll and Mayhew (1976).

Chao (1981) reported that in *Paspalum conjugatum* the two sperms move to the tip of the degenerated synergid and become spherical. The sperm nucleus is surrounded by a cytoplasmic sheath. The sperms leave the synergid through a small opening in the tip of the synergid wall and enter the central cell leaving behind two small darkly stained bodies.

Cass (1973), in a Nomarski-interference probe of isolated living sperms of barley, has shown that they do not exhibit any directional motility. The discharge of the sperm within a distance of 10–15 mm which is equal to 1–1.5 times the diameter of the sperm, also supports the argument about the lack of motility. Microtubules are described in the sperm cells of *Hordeum* (Cass, 1973) and they are believed to cause change in cell shape rather than bringing about locomotion.

Jensen (1973) is of the opinion that of the two sperms released in the degenerated synergid, one comes in contact with the plasmamembrane of the egg cell and the other comes in contact with the plasmamembrane of the central cell. The plasmamembrane at these fusion points dissolve and the sperm nuclei are released in the egg and central cell respectively.

Maze and Lin (1975) reported that in *Stipa elmeri* the two sperms get released one after the other from the synergid by terminal pore.

Time-lag between Pollination and Fertilization

Syngamy and triple fusion usually begin simultaneously; whereever these processes are disjunctive, it is mostly the triple fusion which comes off earlier. The actual process of nuclear fusion is also more brisk in the central cell than in the egg as in *Triticum* (Hu, 1964; Batygina, 1974) and *Hordeum* (Luxova, 1967).

Table 1. Interval between pollination, syngamy and triple fusion

Taxon	Investigator	Interval between pollination and syngamy	Interval between pollination and triple fusion
Avena sativa	Brown & Shands (1957)	4 hr	
Hordeum distichum	Luxova (1968)	30–35 min.	6–9 hrs
Triticale (Hexaploid)	Kaltisikes (1973)	10–15 hrs	4–5 hrs
Triticum aestivum	Hoshikawa (1959)	2–4½ hrs	2–5 hrs
Triticum spp.	Batygina (1974)	4–5 hrs	1½–2 hrs

UNUSUAL FEATURES

Polyspermy

Polyspermy is of rare occurrence and it is regarded as an abnormality. The release of more than two sperms in an embryo sac is known as polyspermy. This may result due to several reasons. One reason is sometimes more than one pollen tube enters the embryo sac. The second reason is in some pollen grains supernumerary nuclei are present and when such pollen grain germinate, the pollen tubes on reaching the embryo sac discharges more than two sperms in the embryo sac. Thus polyspermy brings about fertilization of other cells of the same embryo sac in addition to the normal egg cell and secondary nucleus.

In *Pennisetum squamulatum* Sindhe *et al.* (1970) reported the entry of more than one pollen tube into the embryo sac. The gametes which are released by the pollen tubes near the vicinity of the egg apparatus fertilize both the egg, and synergid as well as the polar nuclei. This results in the origin of extra embryo within a single embryo sac.

In barley Cass and Jensen (1970) rarely observed an additional pollen tube entering the persistent synergid and discharging its content, including two sperms. This persistent synergid did not show any morphological change, either in anticipation, or in response to the arrival of the pollen tube.

Fertilization of Synergid

Sindhe *et al.* (1970) reported the fertilization of not only the egg and polars but also the synergid. This is due to the entry of more than one pollen tube into the same embryo sac.

Fertilization and Environmental Factors

Ikeda *et al.* (1959; quoted in Hoshikawa 1960b) experimentally fertilized wheat and barley and estimated the optimum temperature for fertilization as 15–20°C. Hoshikawa (1960b), however, observed that in wheat the pollen tubes reached the embryo sac earlier at 30°C than at 20°C and were delayed further at 10°C. Within the range of 10–30°C the time elapsing between pollination and fertilization was reduced in inverse proportion to temperature. Also, fertilization took longer time in plants supplied with high and low levels of nitrogen than those of medium levels. After the discharge of sperm, the above conditions influenced the movement of sperm more than the fusion of nuclei.

CHAPTER 6

ENDOSPERM

In Angiosperms the development of endosperm has been described under three modes, namely Nuclear, the Cellular and the Helobial (Maheshwari, 1950). The family Poaceae is characterized by the presence of Nuclear type of endosperm. The endosperm is the most important nutritive tissue and it nurishes the developing embryo. The fusion between the secondary nucleus and second male gamete brings about the formation of primary endosperm nucleus, which in turn develops into endosperm.

Normally triple fusion occurs between the polars and sperm. But in *Oryzopsis miliacea* (Maze *et al.*, 1970) and *Oryza* (Wu and Ts'ai, 1965) the sperm first fuses with one polar nucleus and this diploid nucleus fuses with the other polar nucleus. In general the primary endosperm nucleus divides earlier than the zygote and forms endosperm. In several apomictic species the division of the fusion product takes place after the division of the proembryo (Nielsen, 1947). In some apomictic species such as *Cymbopogon nardus, C. martinii* (Choda *et al., 1982), Panicum notatum* (Febulaus, 1991) triple fusion is absent and hence the endosperm develops automomously, directly from the secondary nucleus. In apomictic species the ploidy of the endosperm may vary depending on the number of nuclei involved in fusion with the male gametes. For example it may be triploid or pentaploid as in *Pennisetum orientale* (Choda and Kumar, 1981), pentaploid as in *Setaria leucopila, S. villosissima* (Emery, 1957), *Tripsacum dactyloides* (Farquharson, 1955), and *Paspalum distichum* (Choda and Bhanwra, 1977) or hexaploid as in *Paspalum secans.*

Theories of Endosperm Origin and Development

There are two theories of endosperm origin and development. According to first theory put forth by Brenchley (1909), in wheat the entire endosperm tissue arises by cellularization of syncytial mass formed as a result of multiple free nuclear divisions of the first endosperm nucleus.

The second theory of Gordon (1922) states that the starchy endosperm tissue arises by serial tangential cell divisions of a cambial layer lining the embryo sac. This layer is derived by multiple division of the first endosperm nucleus. This layer is present as aleurone layer at maturity.

Sandstedt (1946), Yampolsky (1957), Jennings and Morton (1963a) and Buttrose (1963) have combined the two theories suggesting that the inner endosperm arises as described by Gordon. The marked difference in appearance between the sub-aleurone layer and the remainder of the endosperm tissue referred to above made the combination of the two theories.

Development of Endosperm

The first division of the primary endosperm nucleus occurs before the zygote divides. At early stages the primary endosperm nucleus undergoes free nuclear divisions and the daughter nuclei are arranged themselves along the periphery of the embryo sac (Fig. 14A-C). The central part of the embryo sac is occupied by a large vacuole which during further development of the endosperm becomes obliterated.

Normally cell walls appear first around the proembryo and progresses towards the chalazal end (Fig. 14D-F). However there are two exceptions where the cellularization occurs first near the antipodals and seed coat. In *Zizania aquatica* (Weir and Dale, 1960) cell walls form first around the antipodals whereas in *Eragrostis ciliaensis* (Stover, 1937) wall formation takes place first near the seed coat.

Zea mays, after a phase of free-nuclear division throughout the endosperm, mitotic activity is retained by only the peripheral cells after 16 to18 days after pollination. At the end of mitotic divisions, the endosperm nuclei undergo chromosome endoreduplication attaining high ploidy levels (384n) (Tschermak-Woess and Enzenberg-Kunz, 1965). It was observed by Duncan and Ross (1950) that nuclei in the adgerminal layer of surface cells were generally smaller than those on the abgerminal layer.

Cellularization of the Endosperm

Cellularization of the endosperm takes place at various stages of embryo development and it varies from species to species. For example cellularization occurs at 4-celled proembryo stage in *Digitaria bicornis*, *Chloris roxburghiana* (Febulaus and Pullaiah, 1991), 7-celled stage in *Oryzopsis miliacea*, 8-celled stage *Brachiaria reptans,* 16–32 celled stage in *Eragrostis viscosa* (Febulaus and Pullaiah, 1991), 30-celled stage in *Stipa lemmoni* (Maze *et al.*, 1972) and 40-celled stage in *Stipa tortilis* (Maze *et al.*, 1970). In most of the members endosperm becomes completely cellular at globular stage or later stages of embryo development. But in *Melanocenchris jacquemontii* the endosperm becomes completely cellular even at 4-celled proembryo stage (Febulaus and Pullaiah, 1991). The endosperm cells nearest the embryo and antipodals are reported to have densest cytoplasm whereas the rest of the endosperm cells have vacuolate cytoplasm (Maze and Bohm, 1977).

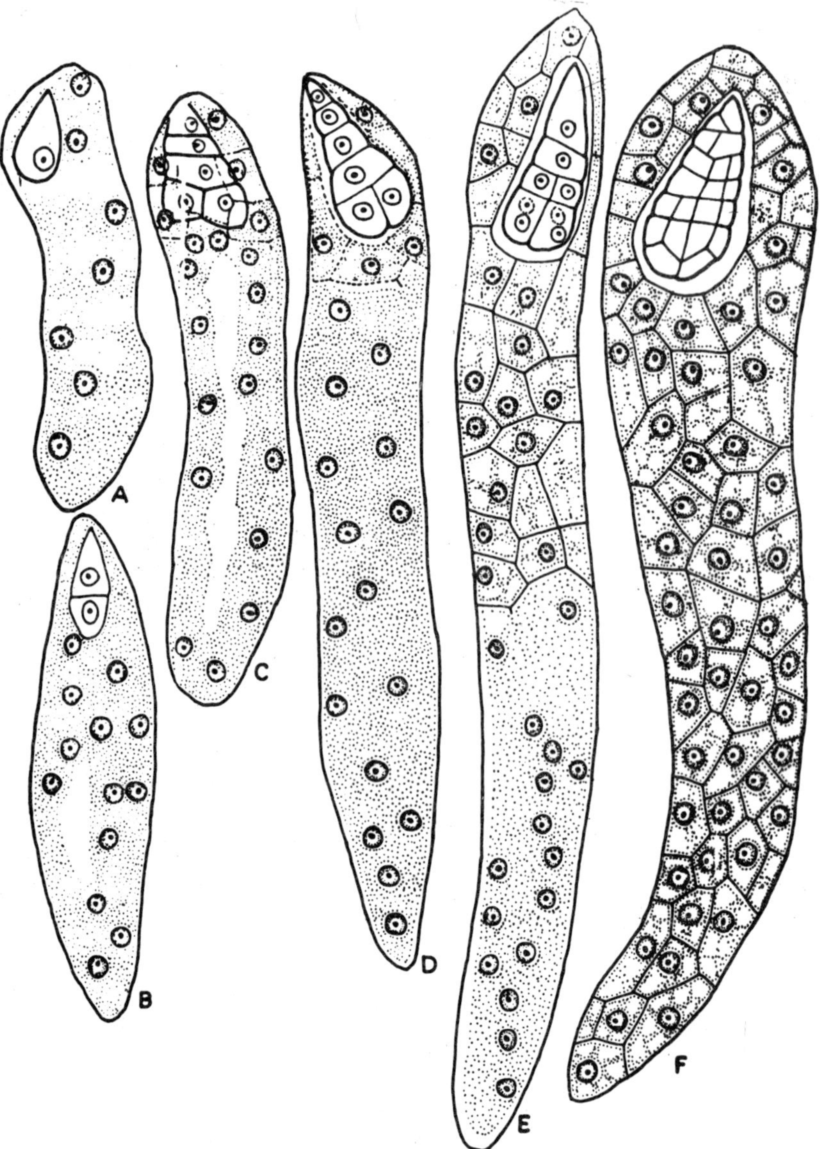

Fig. 14. Stages in the development of endosperm in *Chloris roxburghiana* (Febulaus & Pullaiah, 1991a).

The process of cellularization in wheat has been studied by Buttrose (1963), Evers (1970), Mares *et al.* (1975, 1977), Morrison and O'Brien (1976) and Fineran *et al.* (1982). Three different opinions were made by them on the mode of formation of walls in the free nuclear endosperm.

According to Morrison and O'Brien (1976) the process of cellularization starts with the development of centripetally growing walls from the central cell wall. The free ends of these walls branch and the branches from the adjacent walls fuse with each other. This results in the formation of single layer of uninucleate cells. Regular anticlinal and periclinal divisions in these cells complete cellularization.

Mares *et al.* (1975, 1977) are of the opinion that the anticlinal walls arise from the central cell wall but they do not branch. They grow independently and form open cylinders. First periclinal walls arise within these cylinders. These open cylinders are gradually cleaved into smaller chambers as a result of phragmoplast following nuclear division. Thus a peripheral layer of cells is formed. The free end of the anticlinal wall grow again and second periclinal division results in the formation of second peripheral layer of cells. This process continues till the anticlinal walls from both sides of the central cell meet each other.

Fineran *et al.* (1982) are of the opinion that both the anticlinal and periclinal walls are associated with the formation of phragmoplast. According to them a typical phragmoplast develops between the free endosperm nuclei that occupied the peripheral cytoplasm near the central cell wall. The anticlinal walls arise through cytokinesis around the nuclei and they fuse with each other. Due to this the nuclei come to lie within the open cylinder. This open cylinder undergoes cell division and form periclinal wall by normal cytokinesis. The inner compartments continue to extend centripetally till they meet the anticlinal walls arising from the opposite side of the central cell.

Duvick (1955) has presented a coherent picture of the developmental changes in the various cytoplasmic inclusions in maize endospherm. At the coenocytic stage and so long as the cells are undifferentiated, the endospherm contains small granules and filamentous inclusions (1μ or less). Some of the filaments in the basal endosperm cells grow to appreciable lengths (25μ). Those in the central cells also enlarge (up to about 5μ) and show knobs, each of which apparently develops into a simple starch grain. The protein granules of the mature endosperm are derived from small granules found in undifferentiated cells. The number and size of the protein granules become larger as one proceeds from the central to the peripheral part of the endosperm.

The Aleurone Layer-aleurone Cells

The peripheral layer of the endosperm shows meristematic activity for sometime. This peripheral layer of the endosperm functions as a cambium and produces a series of thin walled cells towards the inner side of the seed (Maheshwari, 1950). In the subsequent stages as the seed develops, the peripheral layer gets transformed into aleurone layer (Jayalakshmi and Lakshmanan, 1982). Bhanwra (1988) reported that from the globular proembryo stage onwards the outer most layer of the endosperm differentiates into an aleurone layer. In most of the members the aleurone cells remain in single layer. However in some plants like *Oryza* (Juliano and Aldama, 1937) and *Poa* (Anderson, 1927) two or three layers of aleurone cells have been reported. In barley the aleurone tissue constitutes three to four layers.

The cells of the aleurone layer are small and appear rectangular in longitudinal and cross sections. Each cell contains an ovoid nucleus and aleurone grains embedded in its cytoplasm. The cell walls are strongly cutinized (Narayanaswami, 1953). In *Setaria* the aleurone cells near the placental vascular supply are columnar and show wall in growth. They help in transferring substances from the vascular tissue to the embryo and endosperm proper. These cells are known as transfer aleurone cells (Bhojwani and Bhatnagar, 1974). Transfer aleurone cells have also been reported in *Echinochloa* and *Zea mays* (Bhojwani and Bhatnagar, 1974), *Chloris barbata, Leptochloa chinensis, Tripogon filiformis, Polypogon fugax, P. monspeliensis, Sporobolus diander* and in *Tragus biflorus* (Bhanwra *et al.*, 1981).

In rice two different types of aleurone cells are reported: (i) the cells surrounding the embryo are rectangular with less dense cytoplasm and (ii) those around the starchy endosperm are cuboidal with dense cytoplasm. Due to their different structure the two types of aleurone cells may have different function (Bechtel and Pomeranz, 1977).

The aleurone cells are characterized by the presence of thick walls without any vacuoles. They are interconnected by plasmodesmata. The organelles of these cells are aleurone grains followed by spherosomes. Aleurone grains are surrounded by a single unit membrane which is closely associated with spherosomes. Aleurone grains are rich in protein, phytin, phospholipids and some carbohydrates.

Each of these components is localized in discrete particles (Jacobson *et al.* 1971). Structurally, the aleurone grains possess two kinds of inclusions, besides the ground substance: (a) Globoids, present within the globoidal cavities—they contain phytin and lipids; (b) protein carbohydrate bodies, which are 1–1.5 μm in diameter. The ground substance also contains high concentration of protein, but it is less than that in the protein—carbohydrate bodies.

Aleurone tissue has received considerable importance in studies of plant hormone action. During seed germination the reserve food (starch and protein) in the endosperm cells is digested by the activity of certain hydrolytic enzymes (amylase and protease) which are secreted by the aleurone cells. Gibberillins have been shown to activate these enzymes or induce their de novo synthesis.

Corn flour, when finely ground, shows a wide variation in granulation, and size and content of endosperm tissue. A general outline of corn kernel shows hull and bran layers, embryonic axis (germ), scutellum, and various endosperm layers. A good understanding of the kernal structure, especially the endosperm tissue, is important so that there is minimum damage to the grain during harvest, drying, storage and marketing (Bechtel and Pomeranz, 1978; Pomeranz and Bechtel, 1978). Scanning electron microscopic studies reveal that starch grains of wheat vary considerably in shape and surface structure, and also change during sprouting. Grains sprouted for 2 days show enzymatic degeneration of starch granules near the aleurone layer. Most of the activity is confined to the larger (A type) granules rather than to the smaller (B-type) ones, suggesting that the former granules differ in physical properties. Such studies on degeneration of starch by enzymes are necessary for understanding the damage caused to the flour. Sprouted wheat-flour affect considerably bread-making quality (Dronzek *et al.* 1972).

SPECIAL FEATURES

i) Endosperm Haustoria

Formation of endosperm haustoria are of rare occurrence in Poaceae. One such report was by Jayalakshmi and Lakshmanan (1980) in *Spinifex littoreus.* According to them the chalazal haustorium arises in the central region of the endosperm, develops as a small pouch with protrusion in the middle. However the exact origin and its haustorial nature are not known (Jayalaskhmi and Lakshmanan, 1980).

ii) Endosperm Nodules

Formation of endosperm nodules have been reported in *Pennisetum typhoideum* (Singh, 1955). These endosperm nodules are rich in various organelles and they are enucleate. The exact function of these nodules are not yet ascertained.

iii) Liquid or Soft Endosperm

Herz (1895), for the first time reported the occurrence of liquid or soft endosperm in grasses. Matlakowna (1913) described the physical and chemical characteristics of endosperm in 20 species belonging to 11 genera. Martin (1946) stated that seven grass genera have soft, fleshy seeds, but he did not mention degrees of softness.

Brown (1955) reported the occurrence of liquid endosperm in *Limnodes arkansana.* He described its characters and stated that it remains fluid for many years. A 15-year old herbarium specimen also showed the same condition. Later Dore (1956) listed 18 species of *Helictotrichon, Koeleria, Sphenopholis* and *Trisetum* as having liquid endosperm and also discussed systematic relationships and endosperm characteristics. Musi (1963) also reported the occurrence of soft or semi-fluid seeds in number of grasses.

Terrell (1971) surveyed 169 genera of grasses and stated that the endosperms condition ranges from liquid to very hard, having all degrees of softness, and hardness. According to him out of 169 genera, 30 genera have liquid or soft endosperm, 9 having semi-solid endosperm and the remaining are having solid endosperm.

Liquid or soft endosperm has been found only in members of the subfamily Festucoideae (Terrell, 1971). The grains with liquid endosperm have tough seed coat and are enclosed within the spikelet structure till germination. They are full of viscuous whitish endosperm. It is insoluable in water and consists of a colloidal mass of starch grains and oil droplets with considerable water holding ability (Brown, 1955).

iv) Mosaic Endosperm

Lack of uniformity in the tissues of the endosperm results in mosaic endosperm. In *Zea mays* two patches of different colours have been reported. This results in irregular mosaic pattern, a part of the endosperm is starchy and part of the endosperm is sugary (Maheshwari, 1950). Earlier, Webber (1900) also observed red and white colours in *Zea mays* endosperm.

Webber proposed two hypothesis to explain the occurrence of mosaic endosperm. According to first hypothesis in some cases the second sperm nucleus enters the embryo sac but fails to unite with the two polars. However both the second sperm nucleus and the secondary nucleus divide separately resulting in the formation of two distinct groups of nuclei in the protoplasm of the embryo sac. During free nuclear endosperm stage these two groups of nuclei become interspersed in the embryo sac and consequently mosaic endosperm is formed (Maheshwari, 1950).

According to second hypothesis, the second sperm nucleus may fuse with only one of the two polar nuclei. After that the repelled polar nucleus and the fertilized polar nucleus divide indpendently giving rise to two groups of nuclei one from a fertilized polar nucleus containing matenal and paternal elements and the other from an unfertilized polar nucleus containing only maternal element (Maheshwari, 1950).

However none of Webber hypothesis are adequately supported. The independent division of the second male nucleus seems impossible (Cf. East, 1913). The endosperm nuclei with deviating chromosone numbers arise because of distriburbed mitosis. Therefore the occurrence of mosaic endosperm may be due to an aberrant behaviour of the chromosomes or due to somatic mutations (Cf. Clark and Coplan, 1940). However the formation of mosaic endosperm has not so far cytologically demonstrated (Maheshwari, 1950).

v) Xenia

The term xenia was first coined by Focke (1881). It implies the direct or immediate effect of pollen in the character of the seed or fruit. For example certain races of *Zea mays* have yellow (dominant) endosperm while others have white (recessive) endosperm. If pollen from the yellow endosperm race is placed on the stigmas of the white endosperm race the hybrid embryo would show the dominant character of the yellow endosperm when it grows into a plant and fruits in the following season. However the yellow colour appears in the endosperm of the same ovule (Maheshwari, 1950).

CHAPTER 7

EMBRYO

The development of grass embryo is of special interest to the embryologist. The family Poaceae is unique in one aspect of embryology i.e. embryo development. According to Johansen's (1950) system, the development of embryo in the majority of the members conforms to the Poa variation of Asterad type (Venkateswarlu and Devi, 1964; Deshpande, 1976; Gawali, 1977; Kulkarni and Dnyanasagar, 1984). According to Soueges system the development of embryo comes under Megarchetype II, series A and subseries A_2 in the first period (Crete, 1963).

Bruns (1892) gave an account of the structure of embryos in 60 species. Van Tieghem (1897) pointed out two basic differences between the Panicoid and Festucoid types of embyros. Reeder (1957; 1962) investigated 300 species of grasses representing over 150 genera from all the tribes of the family and he recognised four important characters: (i) the course of the vascular system, (ii) the epiblast (whether present or absent), (iii) the lower part of the scutellum (free from coleorhiza or fused with it, and (iv) the cross section of embryonic leaf. Based on these characters, grass embryo conforms to two basic types; the Festucoid and the Panicoid. According to Reeder (1957) in Festucoid type, the lower part of the scutellum is lacking or fused to the coleorhiza and the coleoptile is inserted at approximately at the joint of divergence of the scutellum bundle. In Panicoid embryos a distinct cleft is visible between the lower part of the scutellum and the coleorhiza.

Later the grass embryo is critically studied by Brown (1959; 1960; 1965). According to him the scutellum, coleoptile, mesocotyl and epiblast are not homologous with foliage leaves and typical internodes, but are structures peculiar to grass embryo. The outstanding contribution in respect of embryogeny is that of Guignard (1961, 1962, 1963) which embraces 30 species under many genera. This investigation indicates that the development of embryo conforms to Period I, series A_2 and Megarchetype II. Guignard (1961) supported Reeder's contention (1953a; 1956s) that the coleoptile is the modified first leaf of the vegetative shoot apex.

Table III shows details of embryogeny of the family Poaceae.

Development

The zygote undergoes a short period of rest before the first division. The zygote is surrounded by a thin layer of central cell cytoplasm except at its micropylar end where it is attached to the nucellar cells. Its large nucleus contains diffuse chromatin. The wall fo the zygote is thicker at the micropylar region. Plasmodesmata connections are absent between the zygote and central cell. Mitochondria and plastids are concentrated in the perinuclear cytoplasm. Osmophilic, nonmembrane bounded bodies are randomly distributed throughout the cytoplasm. Endoplasmic reticulum is sparse. Dictyosomes and ribosomes are present. Vacuoles are present in the periphery of the cytoplasm (Norstog, 1972).

The zygote after undergoing a period of rest divides transversely to form a terminal cell *ca* and a basal cell *cb* (Fig. 15A,B). The terminal cell *ca* normally divides vertically and the basal cell *cb* divides transversely resulting in the formation of cells *m* and *ci* (Fig. 15C). However, in *Avena fatua* Cannon (1900) described an abnormal case of vertical division in the zygote. In *Poa pratensis* (Nishimura, 1922 a,b), *Pennisetum typhoideum* (Rangaswami, 1935), *Eleusine coracana* (Krishnaswami and Ayyangar, 1937) and *Spinifex littoreus* (Lakshmanan and Jayalakshmi, 1980), it is reported that the terminal cell of the bicelled proembryo divides transversely and not vertically as is normal for the family. These observations need confirmity as the development of embryo in majority of grasses is described as Asterad type in which the terminal cell *ca* divides vertically.

Normally both the cells *ca* and *cb* divide at a time but sometimes *cb* divides earlier than *ca* as in *Cenchrus ciliaris* (Febulaus and Pullaiah, 1991) resulting in a 3-celled linear proembryo.

The two juxtaposed cells of the tier *ca* undergo one more vertical division at right angles to the previous one producing quadrant *q* (Fig. 15C–F). At the same time cell *m* also divides vertically and cell *ci* divides transversely forming two superposed cells *n* and *n'* (Figs. 15D, 16D). The later divides transversely giving rise to the cells *o* and *p*. The cell *m* undergoes another vertical division.

All the cells of *ca* divide transversely to form an octant with tiers *l* and *l'*. Further vertical and transverse division in tiers *l*, *l'*, *m* and *n* are irregular leading to the formation of globular embryo (Figs. 15E-J, 16C-H). Further divisions in the globular embryo leads to the formation of mature embryo.

The above description is given according to Johansen's (1950) system and the relation of the individual cells of the proembryo to the organs of the mature embryo is shown in a schematic representation as follows:

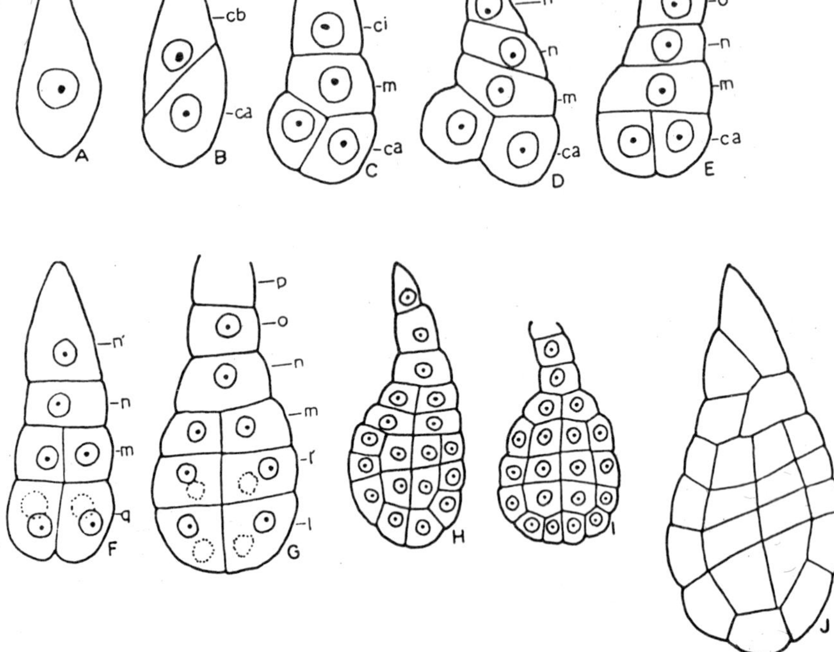

Fig. 15. A-J. Stages in the development of embryo. A-G, J. *Chloris roxburghiana* (Febulaus & Pullaiah, 1991a). H,I. *Eragrostis viscosa* (Febulaus & Pullaiah, 1990).

i) Schematic Representation of the Embryo Development

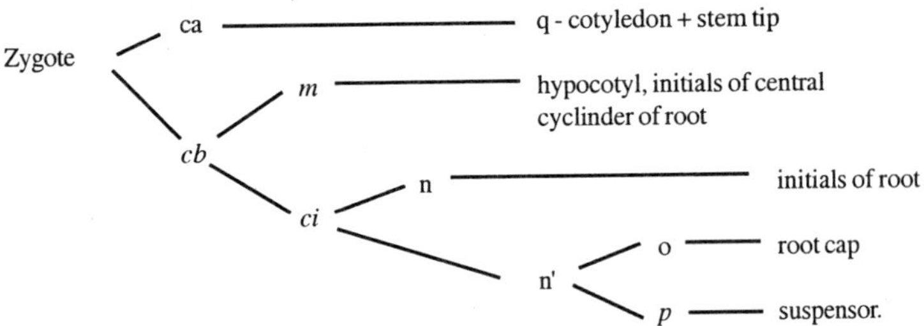

ii) Destination of the individual tiers and the development of embryo

1. First cell generation

Proembryo consists of two cells disposed in two tiers

ca = pco + pvt

cb = phy + icc + iec + co + s.

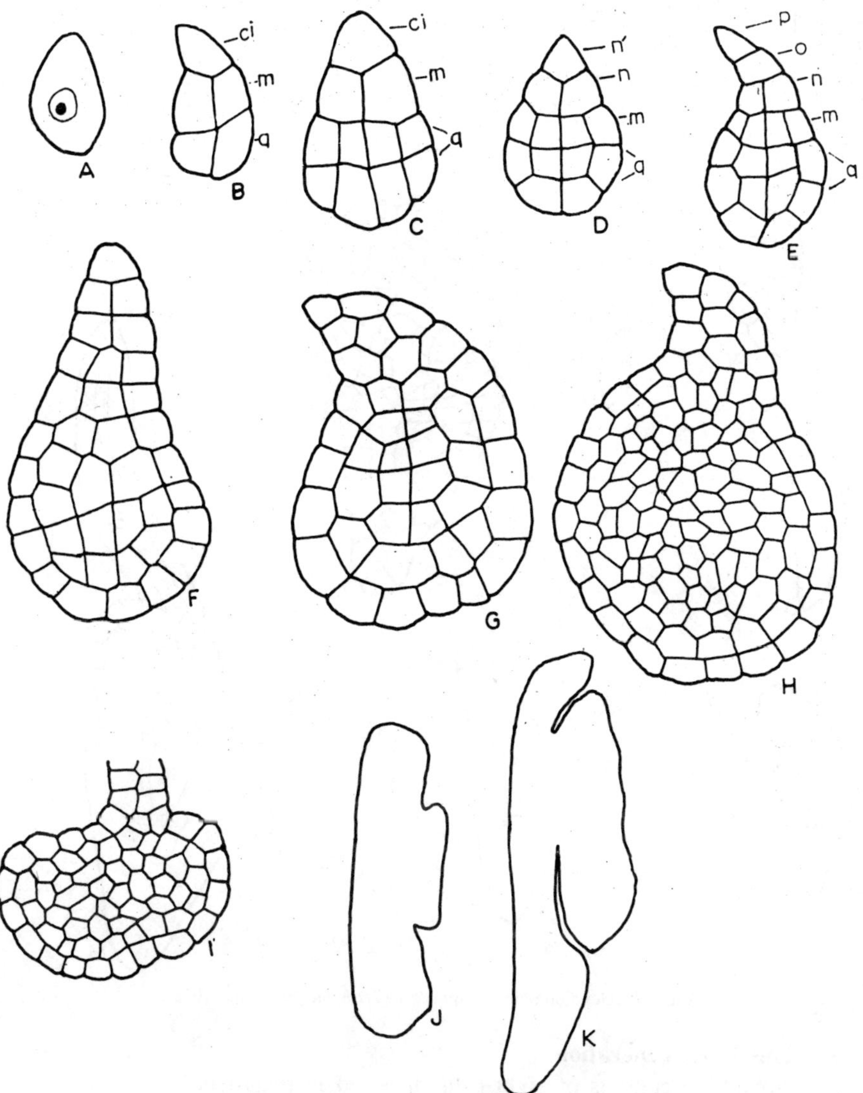

Fig. 16. Stages in parthengenetic embryo development in *Panicum notatum* (Febulaus, 1992).

2. Second cell generation

Proembryo consists of four cells disposed in three tiers.

q = pco + pvt
m = phy + icc
ci = iec + co + s.

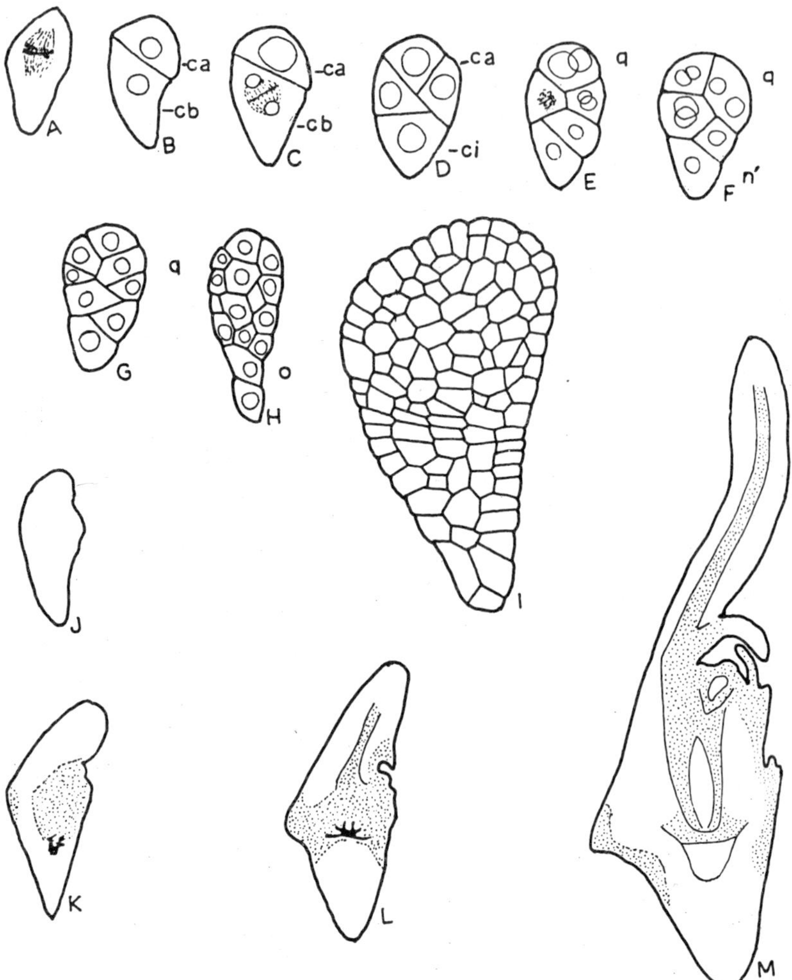

Fig. 17. Development of embryo in *Triticum* (Batygina, 1969).

3. **Third cell generation**

Proembryo consists of eight cells disposed in four tiers.

q = pco + pvt	*n* = iec
m = phy + icc	*n*' = co + s.

4. **Fourth cell generation**

Proembryo consists of sixteen cells disposed in five tiers.

q = pco + pvt
m = phy + iec
n = iec,
o = co
p = s.

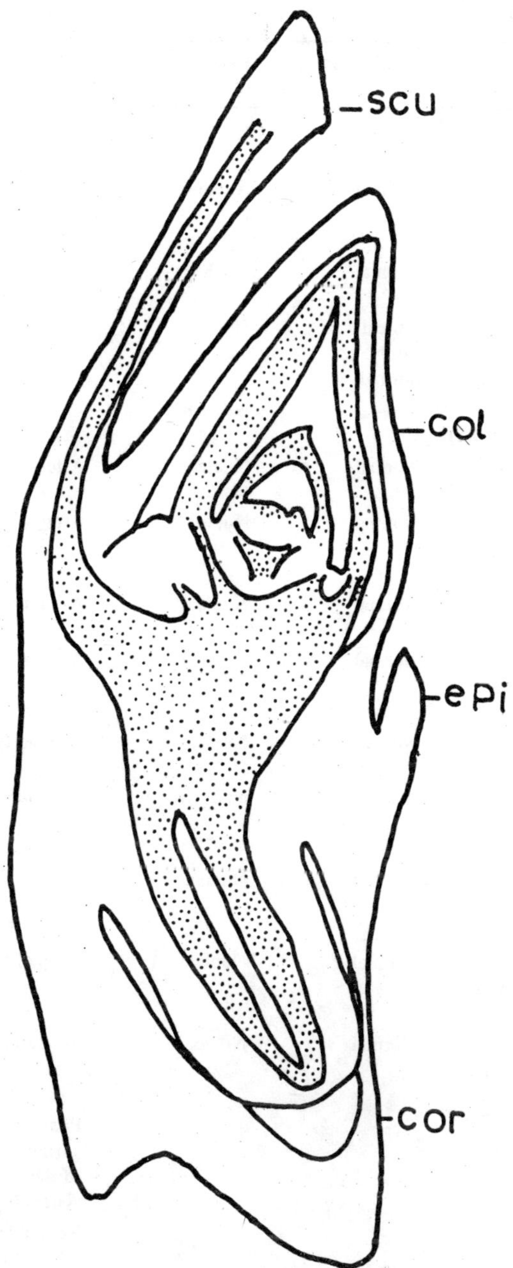

Fig. 18. Mature embryo of *Triticum* as seen in median longitudinal section (Batygina, 1969). scu: scutellum, col: coleoptile, epi: epiblast, cor: coleorrhiza.

Tabel III. Fertilization, Endosperm, Embryo and Polyembryony

Species investigated author	Fertilization	Endosperm development	Embryo development	Polyembryony
Acrachne racemosa (Bhanwra, 1988)	Normal	Nuclear	Asterad type	—
Acroceras munroanum (Shobha & Sindhe, 1992)	Normal	Nuclear	Resembles Asterad type	—
Agrostis pilosula (Muniyamma, 1976)	Normal	Nuclear	Resembles Asterad type	—
Alloteropsis cimicina (Venkateswarlu & Devi, 196	Absent, rarely normal	Nuclear	Resembles Asterad type	—
Apluda mutica (Bhanwra & Pathak, 1985)	—	Nuclear	—	—
Aristida adscensionis (Bhanwra *et al.*, 1988)	Normal	Nuclear	Asterad type	—
Aristida funiculata (Febulaus & Pullaiah, 1993 b)	Normal	Nuclear	Astered type	—
Aristida hystrix (Febulaus & Pullaiah, 1993 b)	Normal	Nuclear	Asterad type	—
Aristida mutabilis (Febulaus & Pullaiah, 1993 b)	Normal	Nuclear	Asterad type	—
Aristida setacea (Febulaus & Pullaiah, 1993 b)	Normal	Nuclear	Asterad type	—
Arundinella mesophylla (Basappa & Muniyamma, 1981)	Normal	Nuclear	Asterad type	—
A. nepalensis (Bhanwra *et al.*, 1982; Bhanwra, 1988)	Normal	Nuclear	Asterad type	—
A. purpurea (Basappa & Muniyamma, 1981)	Normal	Nuclear	Asterad type	—
Arundo donax (Bhanwra *et al.*, 1982; Bharwra, 1988)	Normal	Nuclear	Asterad type	—
Avena fatua (Sharma *et al.*, 1991)	—	—	Poa variation of the Asterad type	—
Bambusa arundinacea (Philip, 1972)	—	—	Resembles Asterad type	—

Species investigated author	Fertilization	Endosperm development	Embryo development	Polyembryony
Brachiaria distachya (Bhanwra *et al.*, 1982; Bhanwra, 1988)	Normal	Nuclear	—	Nucellar embryos
B. eruciformis (Febulaus & Pullaiah 1998)	Normal	Nuclear	Asterad type	—
B. ramosa (Bhanwra *et al.*, 1986)	Normal	Nuclear	—	—
B. remota (Venkateswarlu & Devi, 1964)	Normal	Nuclear	Asterad type	—
B. reptans (Febulaus & Pullaiah, 1991 b)	Normal	Nuclear	Asterad type	—
Bromus inermis (Dzevaltovski and Shpilevaya, 1978)	—	—	—	Adventive embryony
Calamogrostis chalybaea (Stenar, 1932; Nygren, 1946)	Absent	Autonomous	Parthenogenic	Polyembryonate
C. perplexa (Greene, 1984)	Normal	Autonomous	—	—
C. pickeringii (Greene, 1984)	Absent	Nuclear	Normal	—
C. purpurea (Nygren, 1948)	Absent	Autonomous	Parthenogenic	Polyembryonate
C. stricta (Greene, 1984)	Absent	Autonomous	Autonomous	—
Capillipedium huegelli (Choda & Bhanwra, 1980; Bhanwra, 1988)	Absent	Autonomous	—	—
C. parviflorum (Celarier & Harlan, 1957; Bhanwra *et al.*, 1982; Bhanwara, 1988).	Absent	Nuclear	Parthenogenic	—
Cenchrus biflorus (Bhanwra *et al.*, 1982; Bhanwra, 1988)	—	Nuclear	—	—
Cenchrus ciliaris (Snyder *et al.*, 1955;	Absent	Nuclear	Parthenogenic Asterad	Cleavage or adventive

(contd.)

(Table III contd.)

Species investigated author	Fertilization	Endosperm development	Embryo development	Polyembryony
Gupta & Yasvir, 1971; Febulaus & Pullaiah, 1995)			type	polyembryony
C. glaucus (Fisher *et al.*, 1984; Shanthamma, 1982)	Absent Pseudogamous	Nuclear	Autonomous	Polyembryonate
Chloris barbata (Bhanwra *et al.*, 1981)	Normal	—	—	—
Chloris gayana (Chikkannaiah & Mahalingappa, 1976)	—	Nuclear type	Asterad type	—
C. roxburghiana (Febulaus & Pullaiah, 1991a)	Normal	Nuclear	Asterad type	—
Cortaderia jubata (Phillipson, 1978a)	—	—	—	Nu cellar
Cymbopogon caesius (Sreenivasa Rao, 1997)	—	Nuclear type	—	—
Cymbopogon coloratus (Sreenivasa Rao, 1997)	—	"	—	—
Cymbopogon flexuosus (Choda *et al.*, 1982)	—	Nuclear	—	—
C. perkeri (Choda *et al.*, 1982)	—	Nuclear	—	—
Dendrocalamus hamiltonii (Harigopal & Mansai Ram, 1981)	—	Nuclear	—	—
Dendrocalamus longispathus (Harigopal & Mohan Ram, 1987)	—	Nuclear	—	—
D. intermedium (Saran & de Wet, 1970)	Normal	Nuclear	Asterad type	—
Digitaria adscendens (Kulkarni & Dnyansagar, 1981)	—	Nuclear	Asterad type	—
Digitaria bicornis (Venkateswarlu & Devi, 1964; Febulaus & Pullaiah, 1996)	Normal	Nuclear	Asterad type	—
D. biformis (Venkateswarlu & Devi, 1964)	Normal	Nuclear	Asterad type	Cleavage polyembryony
D. ciliaris (Febulaus & Pullaiah, 1996)	Normal	Nuclear	Asterad type	—
Echinochloa colonum (Bhanwra & Choda, 1986)	Normal	Nuclear	Asterad type	—

Species investigated author	Fertilization	Endosperm development	Embryo development	Polyembryony
E. crusgalli (Bhanwra & Choda, 1986)	Normal	Nuclear	Asterad type	—
E. frumentacea (Narayanaswami, 1955b)	Normal	Nuclear	Asterad type	—
E. stagnina (Muniyamma, 1978)	Normal	Nuclear	Asterad type	—
Eleusine coracana (Ayyangar, 1930, Narayanaswami, 1930)	Normal	Nuclear	—	Twin embryos
E. indica (Chandra, 1963, Bhanwra, 1988)	Normal	Nuclear	Asterad type	—
Elymus rectisetus (Crane & Carman, 1957; Hair, 1956)	—	—	—	Polyembryony
Eragrostiella bifaria (Venkateswarlu & Devi, 1964; Febulaus & Pullaiah, 1992b)	Normal	Nuclear	Asterad type	Nucellar polyembryony
Eragrostis bicolor (Streetman, 1963)	Absent	Nuclear	Asterad type	—
E. ciliaris (Bhanwra, 1986a)	Absent	Nuclear	Asterad type	—
E. ciliensis (Kulkarni & Dnyansagar, 1984)	Absent	Nuclear	Asterad type	—
E. coarctata (Venkateswarlu & Devi, 1964; Chandra, 1976)	Absent	Nuclear	Asterad type	—
E. curvula (Streetman, 1963)	Absent	Nuclear	Pseudogamous	—
E. diarrhena (Venkateswarlu & Devi, 1964)	Absent	Nuclear	Asterad type	—
E. intermedia (Streetman, 1963)	Absent	—	Pseudogamous	—
E. lehmanniana (Streetman, 1963)	Absent	—	Pseudogamous	—
E. namaquensis (Ghaisas & Patil, 1990)	Normal	Nuclear	Asterad type	—
E. nigra (Bhanwra, 1986a)	Normal	Nuclear	Asterad type	—
E. pilosa (Chandra, 1976)	Normal	Nuclear	Asterad type	—
E. plano (Streetman, 1963)	Absent	—	Pseudogamous	—

(contd.)

(Table III contd.)

Species investigated author	Fertilization	Endosperm development	Embryo development	Polyembryony
E. plumosa (Venkateswarlu & Devi, 1964)	Normal	Nuclear	Asterad type	—
E. superba (Streetman, 1963)	Absent	—	Pseudo-gamous	—
E. tef (Longly *et al.*, 1985)	Normal	Nuclear	—	—
E. tenella (Venkateswarlu & Devi, 1964; Kalantri, 1990)	Normal	Nuclear	Asterad type	—
E. tremula (Bhanwra, 1986a)	Normal	Nuclear	Asterad type	—
E. unioloides (Chandra, 1976)	Normal	Nuclear	—	—
E. viscosa (Venkateswarlu & Devi, 1964; Febulaus & Pullaiah, 1990)	Normal	Nuclear	Asterad type	—
Eriochloa procera (Venkateswarlu & Devi, 1964; Shanthamma & Narayan, 1976–7)	Normal	Nuclear	Asterad type	polyembryonate
Euchlaena mexicana (Koul & Sahi, 1962)	Normal	Nuclear	—	—
Glyceria tonglensis (Bhanwra, 1988)	Normal	Nuclear	Asterad type	—
Helictotrichon virescens (Bhanwra, 1988)	Normal	Nuclear	—	—
Hordeum vulgare (Cass & Peteya, 1985)	Normal	Nuclear	Asterad type	—
Imperata cylindrica (Bhanwra, 1988)	Normal	Nuclear	Asterad type	—
Ischaemum rugosum (Sreenivasa Rao, 1997)	—	Nuclear type	Poa variation of Asterad type	—
Iseilema anthephoroides (Venkateswarlu & Devi, 1964)	Normal	Nuclear	Asterad type	—
I. prostratum (Bhanwra *et al.*, 1982)	Normal	Nuclear	Asterad type	—
Jansenella griffithiana (Shobha & Sindhe, 1981, 1987)	—	Nuclear	Resembles Asterad type	—
Leptochloa chinensis (Bhanwra *et al.*, 1981)	Normal	Nuclear	Asterad type	—

Species investigated author	Fertilization	Endosperm development	Embryo development	Polyembryony
L. neesii (Venkateswarlulu & Devi, 1964)	Normal	Nuclear	Asterad type	—
L. panicea (Venkateswarlu & Devi, 1864; Bhanwra, 1988)	Normal	Nuclear	Asterad type	—
Limnopoa meeboldii (Shobha & Sindhe, 1996)	—	Nuclear	Resemblese Asterad type	—
Melocalamus compactiflorus (Harigopal & Mohan Ram, 1987)	—	Nuclear	—	—
Ochlandra travancorica (Harigopal & Mohan Ram, 1987)	—	Nuclear	—	—
Oplismenus compositus (Bhanwra & Soni, 1986)	—	—	—	—
O. burmanii (Bhanwra & Soni, 1986)	—	—	—	—
Oropetium thomaeum (Diwanji & Diwanji, 1981)	—	Nuclear	Asterad type	—
O. vellosulum (Diwanji & Diwanji, 1981)	—	Nuclear	Asterad type	—
Oryza alta (Sreenivasa Rao, 1997)	Normal	Nuclear	Asterad	—
Oryza brachyantha (Sreenivasa Rao, 1997)	"	"	"	—
Oryza latifolia (Venkateswarlu & Devi, 1964)	"	"	"	—
Oryza longistaminata (Sreenivasa Rao, 1997)	"	"	"	—
Oryza malampuzhaensis (Sreenivasa Rao, 1997)	"	"	"	—
Oryza minuta (Sreenivasa Rao, 1997)	"	"	"	—
Oryza officinalis (Sreenivasa Rao, 1997)	"	"	"	—
Oryza rhizomatis (Sreenivasa Rao, 1997)	"	"	"	—
Oryza ridleyi (Sreenivasa Rao, 1997)	"	"	"	—
Oryza rufipogon (Sreenivasa Rao, 1997)	"	"	"	—
Oryza sativa (Raju, 1980; Sreenivasa Rao, 1997)	Normal	Nuclear	Asterad	—

(contd.)

(Table III contd.)

Species investigated author	Fertilization	Endosperm development	Embryo development	Polyembryony
Panicum antidotale (Warmke, 1954; Shamakumari, 1960)	—	Nuclear	Asterad type	—
P. miliaceum (Khosla, 1946; Narayanaswami, 1955a)	Normal	Nuclear	Asterad type	—
P. miliare (Narayanaswami, 1955a)	Normal	Nuclear	Asterad type	—
P. notatum (Febulaus, 1992)	Absent	Autonomous Nuclear	Parthenogenic Asterad type	—
P. repens (Febulaus & Pullaiah, 1992a)	Normal	Nuclear	Asterad type	—
Paspalidium flavidum (Bhanwra *et al.*, 1982)	Normal	Nuclear	Asterad type	—
Paspalidium punctaum (Gawali, 1990)	—	—	Asterad type	—
Paspalum scrobiculatum (Narayanaswami, 1954)	Normal	Nuclear	—	Antipodal Polyembryony
Pennisetum mezianum (Shanthamma & Naryan, 1977)	Normal	Nuclear	—	Adventive polyembryony
P. oreientale (Choda *et al.*, 1981)	Absent	—	Parthenogenic	—
P. squamulatum (Sindhe *et al.*, 1980)	Absent	Nuclear	—	Nucellar polyembrony
P. typhoideum (Narayanaswami, 1953; Raju, 1980)	Normal	Nuclear	Asterad type	—
Perotis hordeiformis (Venkateswarlu & Devi, 1964; Bhanwra, 1988)	Normal	Nuclear	Asterad type	—
Phalaris minor (Bhanwra, 1988)	Normal	Nuclear	Asterad	—
Poa alpina (Hakansson, 1943)	Absent	Nuclear	Asterad	—
P. annua (Bhanwra *et al.*, 1985)	Normal	Nuclear	Asterad type	—
P. arctica (Engelbert, 1941)	Absent	Nuclear	Parthenogenic	—
P. compressa (Anderson, 1923)	—	—	—	Polyembryony due to multiple embryo sacs

Species investigated author	Fertilization	Endosperm development	Embryo development	Polyembryony
P. pratensis (Nishimura, 1922; Anderson, 1923; Tinney, 1940; Akerberg, 1942, 1943)	Absent	—	Parthenogenic	Twin embryos from twin embryo sacs
Polypogon fugax (Bhanwra *et al.*, 1981)	Normal	Nuclear	—	—
P. monspeliensis (Bhanwra *et al.*, 1981)	Normal	Nuclear	—	—
Pseudostachyum polymorphum (Harigopal & Mohan Ram 1987)	—	Nuclear	—	—
Rhynchelytrum villosum (Shanthamma & Naryan, 1976–77)	—	—	—	Polyembryonate
Rostraria phleoides (Bhanwra, 1988)	Normal	Nuclear	—	—
Rottboellia exaltata (Sreenivasa Rao, 1997)	—	Nuclear	Poa variation of Asterad type	—
Saccharum benghalense (Bhanwra, *et al.*, 1982)	Normal	Nuclear	Asterad type	—
Setaria glauca (Bhanwra *et al.*, 1982)	Normal	Nuclear	Asterad type	—
S. intermedia (Bhanwra *et al.*, 1982)	Normal	Nuclear	Asterad type	—
Setaria interrupta (Sreenivasa Rao, 1997)	—	Nuclear	Asterad type	—
S. italica (Khosla, 1946; Narayanaswami, 1953)	—	Nuclear	—	—
Setaria pumila (Sreenivasa Rao, 1997)	—	Nuclear	Asterad type	—
S. verticillata (Bhanwra *et al.*, 1982)	Normal	Nuclear	Asterad type	—
Sorghum arundinaceum (Venkateswarlu & Devi, 1964)	Normal	Nuclear	Asterad type	—
Spinifex littoreus (Jayalakshmi & Lakshman, 1982)	Normal	Nuclear	Asterad type	—
Sporobolus diander (Bhanwra, 1988)	Normal	Nuclear	Asterad type	—
S. fertilis (Bhanwra, 1988)	Normal	Nuclear	Asterad type	—

(contd.)

(Table III contd.)

Species investigated author	Fertilization	Endosperm development	Embryo development	Polyembryony
Sporobolus indicus (Sreenivasa Rao, 1997)	Normal	Nuclear	Poa variation of Asterad type	—
Sporobolus spicatus (Sreenivasa Rao, 1997	Normal	Nuclear	—	—
Sporobolus tremulus (Seshavatharam & Bhaskara Rao, 1983)	—	Nuclear	Poa vaiation of Asterad type	—
Stipa elmeri (Maze & Bohm, 1973)	Normal	Nuclear	—	—
S. lemmonii (Maze & Bohmn 1973)	Normal	Nuclear	—	—
S. hendersonii (Maze & Bohm, 1973)	Normal	Nuclear	—	—
Themeda triandra (Merwe, 1957)	Normal	Nuclear	—	—
Trachys muricata (Venkateswarlu & Devi, 1964)	Normal	Nuclear	Asterad type	—
Tragus biflorus (Chandra, 1975; Bhanwra *et al.*, 1981)	Normal	Nuclear	—	—
Tripogon filiformis (Bhanwra *et al.*, 1981)	Normal	Nuclear	—	—
Triticum aestivum (Bhatnagar & Chandra, 1978; Batygina, 1969)	Normal	Nuclear	Graminad type	—
Triticum dicoccum (Raju, 1980)	Normal	Nuclear	Asterad type	—
T. durum (Raju, 1980)	Normal	Nuclear	Asterad type	—
Urochloa panicoides (Basavaiah & Murthy, 1989)	Normal	Nuclear	Asterad type	—
Vetiveria zizanioides (Bhanwra, 1988)	Normal	Nuclear	Asterad type	—
Zea mays (Kiesselback, 1926	Normal	Nuclear	Asterad type	False poly-embryony
Zingeria trichopoda (Shehata, 1995)	Normal	Nuclear	Asterad type	—

(pco = cotyledon; pvt = stem tip; phy = hypocotyl; icc = initials of central cylinder of root; iec = initials of root cortex; co = root cap and s = suspensor).

Batygina (1969) has described in detail the mode of embryo development in *Triticum*. All the four species of *Triticum* examined by her show a fixed pattern of embryogeny which is so different from that in dicotyledons and other monocotyledons that she remarked: the unique mode of embryogenesis in Gramineae may allow the separation of a new type of Embryogeny-Graminad type. The early part of embryogeny in *Triticum* is characterised by the regular occurrence of oblique divisions.

The first nuclear division of the zygote (Fig. 17A) is followed by laying down of an oblique wall, cutting a small apical cell (*ca*) and a large basal cell (cb; Fig. 17B). Cell *cb* again divides obliquely forming cells *ci* and *m* (Fig. 17 C,D). The upper end of the wall formed during this division connects with the wall separating *ca* and *cb*. The third division occurs in the cell *ca*, in a plane perpendicular to the first division of the zygote. Thus a T-shaped proembryo (4-celled) is formed (Fig. 17D). However, the orientation of the walls are very different from the T-shaped tetrads in either dicotyledons or other monocotyledons. This characterises the wheat embryo.

Cell *ci* divides by a wall at right angles to the wall between *ci* and *m*, resulting in the formation of cells *n* and *n'* (Fig. 17E). Division of the daughter cells of *ca* are in the same plane as the first division in *ca* but at right angles to it, forming the typical quadrant *q* (Fig. 17E,F). The cell *m* divides vertically into two cells (Fig. 17E–H). Further divisions go on in various planes (Fig. 17G–I).

At 16–32-celled stage of the proembryo organogenesis sets in. The first organ to be initiated is the single cotyledon or scutellum. Its differentiation starts with a growth in the apical-lateral region of the proembryo (Fig. 17I) involving sector *q, m* and *n*. With further development a constriction appears opposite the scutellum (in the sector *ca*) demarcating it from the rest of the embryo (Fig. 17J). This is followed by the appearance of the promordia of coleoptile and then the shoot apex close to the notch (Fig. 17K,L). The radicle differentiates endogenously in the central zone of the embryo (Fig. 17L,M). The epicotyl is formed by the terminal tier (*q*). Its apparent lateral position in the mature embryo is a secondary feature. It arises due to active growth of the cotyledon leaving behind the epicotyl.

Generally in majority of the members of monocotyledons both stem tip and cotyledon arise from the one and the same terminal tier (Schnarf, 1929; Johansen, 1950). However Soueges designated different origins to the stem tip and cotyledon. According to Soueges terminal cell gives rise to only cotyledon and stem tip originates from the middle tier. Accordingly Soueges designated a separate Megarchetype I for monocots.

Natesh and Rau (1984) in their observations pointed out that Soueges (see Crete, 1963) included several Monocots in Period I and designated a separate

Megarchetype for them on the erroneous assumption that in all monocotyledons the stem tip was lateral, originating from the middle tier, that only the cotyledon was terminal.

Although Soueges designated two different tiers for stem tip and cotyledon recent studies on a number of monocotyledons by Haccius, Swamy, Lakshmanan and others (see Lakshmanan, 1972) have clearly demonstrated that the epicotyl and cotyledon arise from one and the same tier. Guignard (1975b) also admitted the origin of the stem tip form the terminal tier and accordingly abandoned the Megarchetype I of Soueges system of classification. Natesh and Rau (1984) also proved that the shoot apex of grass embryo also appears to be lateral and both stem tip and cotyledon arise from the terminal proembryonic tier, but due to the over growth of the cotyledon the stem tip comes to occupy a lateral position.

Structure of Mature Grass Embryo

The mature embryo has a single cotyledon called scutellum. The scutellum is a long, shield-like vascularized structure, laterally attached to the axis (Natesh and Rau, 1984). The portion of the embryonal axis below the level of scutellum is the radicle. The radicle bears an apical meristem and root cap at the lower end and is enclosed by coleorhiza (Fig. 18). The portion of the embryonal axis above the level of scútellum is the epicotyl which comprises of shoot apex with some leaf primordia enclosed by coleoptile. In some grasses especially those belonging to Pooideae an erect flap-like outgrowth is present on the side of the coleorhiza. This small out growth is called epiblast (Fig. 18). Epiblast is always absent in the grasses belonging to the sub-family Panicoideae. It is always present among the members of Festucoideae (Reeder, 1957; Bhanwra, 1988).

Reeder (1957) investigated 300 species of grasses belonging to both Festucoideae and Panicoideae. He observed four differences between Festucoid and Panicoid embryos. (i) Among the members of the Panicoideae coleoptile is inserted at some distance above the point of divergence of the scutellum bundle and there is a distinct internode between coleoptile and scutellum. In members of Festucoideae the coleoptile is inserted at about the point where the scutelum bundle diverges, and no internode is present between coleoptile and scutellum. (ii) In Panicoid embryos a distinct cleft is visible between the lower part of the scutellum and the coleorhiza whereas in Festucoid embryos no cleft is evident since lower part of scutellum is absent or fused with the coleorhiza. (iii) Epiblast is always absent in the grasses belonging to Panicoideae and it is always present in the members of Festucoideae. (iv) The 'seeds' of true Panicoids always have relatively large embryos and the portion of endosperm is correspondingly reduced, whereas the true festucoids have always small embryos and consists of large amount of endosperm (Reeder, 1957).

According to Van Tieghem (1897) Malpighi for the first time gave the description of embryos of two grasses—*Avena* and *Triticum* and recognized the structures such as scutellum, epiblast and coleoptile. Later Norner (1881), Bruns (1892), Kennedy (1900), Soueges (1924) and Avery (1930) reviewed certain phases of embryo morphology.

Norner (1881) studied early division of the zygote and proembryos of *Hordeum, Avena, Triticum* and *Secale*. Soueges (1924) reported that there was a definite arrangement of cells and regular sequence of cell divisions in the development of embryo. However, Randolph (1936) while giving complete description of embryogeny of *Zea* stated that there are no definite arrangement of cells and no regular sequence of divisions. He considered temperature relations as an important factor influencing the rate of embryo development. Merry (1941) gave complete description of *Hordeum sativum* from the time of fertilization to maturity of seed.

CHAPTER 8

APOMIXIS

The term apomixis was first coined by Hacke (1893). Apomixis, as defined by Winkler (1908), is a process, in contrast to Amphimixis (sexual reproduction), the sporophyte is produced from the gametophyte without any gametic fusion. Winkler included vivipary and other forms of vegetative propagation such as bulbils, runners, rhizomes etc. under Apomixis. Fagerlind (1940) also included vegetative propagation in Apomixis. But Gustafson (1946, 1947), Maheshwari (1950) and Battaglia (1963) have not included vegetative reproduction and adventive embryony in apomixis. According to Battaglia (1963) apomixis may be defined as the production of sporophyte from gametophyte without sexual fusion.

Maheshwari (1950) and Battaglia (1963) classified apomixis into two types: Recurrent (apomixis in unreduced embryo sacs) and Non-recurrent (apomixis in reduced embryo sacs). Heslop Harrison (1972) does not recognise the non recurrent type as a part of apomixis as it includes the process of meiosis in the formation of reduced embryo sacs.

Some authors (Nogler, 1984) used the term Gametophytic apomixis. According to Battaglia (1992) the term apomixis means loss of mixis and such a loss must be unequivocally referred to the gametophyte. Consequently it follows that "sporophytic apomixis" has no logical meaning and cannot be established. Hence its equivalent term gametophytic apomixis also is rejected.

The credit for the discovery of apomixis in Gramineae is usually given to Muntzing (1933) for his work on *Poa pratensis* and *Poa alpina*, although apomixis in *Poa alpina* had been suggested by Zollikofer (cited by Myers, 1947) and Stenar's (1932) paper on Parthenogenesis, in the genus *Calamogrostis*. However, Muntzing's work based on progeny tests was very critical and well executed and established apomixis in the genus and gave an impetus for further research. Brown and Emery (1958) in their survey of 164 species belonging to 64 genera have reported that 54 species of Panicoideae are apomictic. They have further suggested that apomictic species are more common in the old world Panicoideae, than the new world Panicoideae.

Overall approximately 406 species belonging to nearly 35 families of plants show apomixis, of these more than 250 species are members of Poaceae (see Table IV).

MECHANISM OF APOMIXIS

In grasses the mechanism of apomixis is associated with the formation of an unreduced embryo sac either through apospory or diplospory both in the facultative and obligate apomicts.

1. Apospory

Origin of Aposporic Embryo Sacs

The aposporic embryo sacs may arise from any of the initials of the nucellus adjoining the sexual embryo sacs as in *Pennisetum dubium* (Gildenhuys and Brix, 1959) or the megaspore mother cell as in *Paspalum dilatatum* (Bashaw and Holt, 1958), *Pennisetum ciliare* (Bashaw, 1962). In *Dicanthium annulatum* (Reddy and D'Cruz, 1969), *Cenchrus ciliaris* (Gupta and Yashvir, 1971; Febulaus and Pullaiah, 1991), *Dicanthium pseudoischaemum* (Febulaus and Pullaiah, 1991), the megaspore mother cell degenerates and aposporic initials develop in the vicinity of the degenerating megaspore mother cell. In *Pennisetum* the megaspore mother cell or its products degenerate autonomously even before the origin of aposporic initials (Shanthamma, 1982). Such cases are also reported in *Pennisetum villosum, P. setaceum* (Narayan, 1951), *Paspalum secans* (Snyder, 1957), *Poa granita* (Skalinska, 1959), *Heteropogon contortus* (Emery and Brown, 1958), *Bothriochloa ischaemum* (Brown and Emery, 1957), *Pennisetum ciliare* (Snyder *et al.*, 1955), *Panicum maximum* (Warmke, 1954) and *Bouteloua curtipendula* (Mahamed and Gould, 1966). The aposporic initials are deep seated in the nucellus in *Apluda mutica, Sorghum halepense* and *Panicum repens* (Seshavatharam and Satyamurthy, 1976). In *Pennisetum villosum, P. setaceum, P. orientalis* and *P. pedicillatum* the aposporous embryo sacs originate from the nucellar cells in the vicinity of the megaspore mother cell or its products or in the chalazal region, integument, funiculus and in extreme cases from the pericarpic region (Narayan, 1962). Moskova and Yakovlev (1974) also reported that in *Bothriochloa ischaemum* the aposporous embryo sacs develop from the central region of the nucellus adjoining the sexual embryo sacs or rarely develop either in the funicular region of the ovule or from the cells of the inner integument.

Hakansson (1943) investigated both sexual and apomictic strains of *Poa alpina* where the sexual strain gave rise to a haploid 8-nucleate embryo sac where as the apomictic strain gave rise to a diploid embryo sac. Nielson (1947a) gave developmental sequence of embryo and endosperm in apomictic and sexual forms of *Poa pratensis.*

Asker (1979) refers to four possible modes of reproduction in facultative apomicts: (1) The unreduced cells by parthenogensis produce uniform maternal offsprings; (2) the unreduced egg cells, by fertilisation give rise to autotriploids

Table IV. Apomixis

Species investigated and author	Type of Apomixis	Mechanism of Apomixis
Agropyron scabrum (Hair, 1956)	Apospory	Facultative
Anthephora pubescens (Brown and Emery, 1958)	Apospory	4-nucleate
Anthoxanthum odoratum (Shanthamma and Narayan, 1976–77)	Apospory	Facultative
Apluda mutica (2n = 20, 60, 70) (Murthy, 1973)	Apospory Panicum type Parthenogenetic	Facultative, 4-nucleate
Arundinella mesophylla (Basappa and Muniyamma, 1981)	Apospory Panicum type	Facultative 4-nucleate, rarely 5-nucleate, nonfunctional
Bothriochloa caucasica (Celarier and Harlan, 1957; Harlan *et al.*, 1964)	Apospory Pseudogamous	4-nucleate
B. decipiens (Brown and Emery, 1958)	Apospory Pseudogamous	4-nucleate
Bothriochloa ewartiana (Carman, 1995)	Apospory	Apospory
B. glabra (Celarier and Harlan, 1957)	Apospory Pseudogamous	4-nucleate
Bothriochloa grahamii (Harlan *et al.*, 1964)	Apospory	—
B. insculpta (Celarier and Harlan, 1957)	Apospory Pseudogamous	4-nucleate
B. intermedia (Celarier and Harlan, 1957)	Apospory Pseudogamous	4-nucleate
B. ischaemum (2n = 40) (Brown and Emery, 1957a; Celarier and Harlan, 1957; Christoff and Moskova, 1972)	"	4-nucleate Obligate
B. odorata (2n = 40) (Choda and Bhanwra, 1977; Bhanwra, 1988)	Apospory Pseudogamous	Obligate, 4-nucleate
B. pertusa (2n = 40) (Celarier and Harlan, 1977, Gupta, 1968)	Apospory	Facultative, 4-nucleate
Bothriochloa radicans (Brown and Emery, 1958)	Apospory	—
Bothriochloa serrata (Carman, 1995)	Apospory	—
B. swartziana (Brown and Emery, 1958)	Apospory	4-nucleate

Species investigated and author	Type of Apomixis	Mechanism of Apomixis
B. venusta (Celarier and Harlan, 1957)	Apospory	Obligate
Bouteloua curtipendula (Mohamed and Gould, 1966)	Apospory	Obligate, 4-nucleate
B. gracilis (Snyder and Harlan, 1953)	Gonial apospory Hieracium type	—
Brachiaria bovonei (do Valle and Glienke, 1991)	Apospory	—
Brachiaria brizantha (2n=4X = 36) (Brown and Emery, 1958; Lutts *et al.*, 1991, 1994)	Apospory	Obligate, 4-nucleate
B. decumbens (2n=4X = 36)* (Pritchard, 1967; Lutts *et al.*, 1991, 1994) * Diploids are sexual	Apospory Parthenogenesis	4-nucleate Obligate
B. dictyoneura (doValle and Glienke, 1991)	Aprospory	—
B. humidicola (doValle and Glienke, 1991)	Apospory	—
B. jubata (doValle & Glienke, 1991)	Apospory	—
B. leucacrantha (doValle and Glienke, 1991)	Aposproy	—
B. milliformis (do Valle and Glienke, 1991)	Apospory	—
B. nigropedata (doValle & Glienke, 1991)	Apospory	—
B. platynota (doValle and Glienke, 1991)	Apospory	—
B. serrata (Brown and Emery, 1958)	Apospory	Obligate, 4-nucleate
Brachiaria ruziziensis (Cruz *et al.*, 1989)	Apospory	Obligate
B. setigera (2n = 36) (Muniyamma, 1977)	Pseudogamy	Facultative
B. subulifolia (doValle and Miles, 1992)	Apospory	—
Buchloe sp. (Carman, 1997)	Apospory	—
Calamogrostis brachytricha (Tateoka, 1969)	—	Facultative
C. canadensis (Nygren, 1954; Jorgensen *et al.*, 1958)	Diplospory	Facultative
C. chalybaea (Stenar, 1932; Nygren, 1946)	Diplospory	Obligate, 8-nucleate

(contd.)

(Table IV contd.)

Species investigated and author	Type of Apomixis	Mechanism of Apomixis
C. crassiglumis (Nygren, 1954; Jorgensen *et al.*, 1958)	Diplospory, Embryo develops autonomously	Obligate, 8-nucleate
C. hakonensis (Tateoka 1968, 1973, 1976)	"	"
C. hyperborea (Jorgensen *et al.*, 1958)	—	8-nucleate
C. inexpansa (Greene, 1984)	Diplospory Taraxacum type	8-nucleate
C. lansdorfii (Carman, 1995)	Diplospory	—
C. laponica (Nygren, 1946; Jorgensen *et al.*, 1958)	Diplospory	Facultative, 8-nucleate
C. purpurascens (Nygren, 1954)	Diplospory	Obligate, 8-nucleate
C. purpurea (Nygren, 1946)	Diplospory	Facultative, 8-nucleate
C. sachalinensis (Tateoka, 1973)	Diplospory	Facultative, 8-nucleate
C. stricta (Greene, 1984)	Diplospory Parthenogenic	Facultative, 8-nucleate
Capillipedium huegelli (2n = 40) (Choda and Bhanwra, 1980; Bhanwra, 1988)	Apospory pseudogamous	Facultative, Facultative,
C. parviflorum (Celarier and Harlan, 1957; Bhanwra *et al.*, 1982; Bhanwra, 1988)	Apospory, parthenogenesis & pseudogamous	Obligate, 4-nucleate
C. spicigerum (Celarier and Harlan, 1957)	Apospory, parthenogenesis & pseudogamous	4-nucleate
Cenchrus ciliaris (2n = 45) (= *Pennisetum ciliare*) (Snyder *et al.*, 1955; Gupta and Yashvir, 1971; Shanthamma and Narayan, 1976-77, Febulaus and Pullaiah, 1995; Simpson and Bashaw, 1969)	Apospory Panicum type Diplospory Pseudogamous	Facultative, 4-nucleate
C. glaucus (2n = 45) (Fisher *et al.*, 1954; Shanthamma, 1982)	Apospory Panicum type Pseudogamous	Obligate, 4-nucleate
C. pennisetiformis (Das and Islam, 1977)	Apospory	Facultative, 4-nucleate
C. setigerus (Fisher *et al.*, 1954; Das and Islam, 1977; Bhanwra *et al.*, 1991)	Apospory	Facultative, 4-nucleate
Chionachne koenigii (Seshavatharam, 1983)	Apospory Panicum type Pseudogamous	Facultative, 4-nucleate

Species investigated and author	Type of Apomixis	Mechanism of Apomixis
Chloris andropogonoides (Carman, 1995)	Apospory	—
Chloris cucullata (Carman, 1995)	Apospory	—
Chloris gayana (Brown and Emery, 1958)	Apospory	Facultative, 4-nucleate
Chloris verticillata (Carman, 1995)	Apospory	—
Chrysopogon aciculatus (Seshavatharam, 1983)	Apospory Panicum type, Pseudogamous	Facultative, 4-nucleate
Coix aquatica (2n = 10) (Koul, 1970; Venkateswarlu and M.K. Rao, 1966)	Apospory Panicum type	Facultative, 4-nucleate
Coix gigantea (2n = 20) (Narayan, 1967–68)	Apospory Panicum type	Facultative, 4-nucleate
Coix lachryma-jobi (2n = 20) (Seshavatharam, 1983; Mell *et al.*, 1995)		
Cortaderia atacamensis (Connor, 1974)	Apospory Parthenogensis	—
Cortaderia jubata (Philipson, 1978; Connor and Dawson, 1993))	Apospory	Facultative, 4-nucleate
Cortaderia rudiscula (Philipson; 1978; Connor and Dawson, 1993)	Apospory	—
Cortaderia speciosa (Connor, 1974)	Apospory Parthenogenesis	—
Dicanthium annulatum (2n = 60) (Celarier and Harlan, 1957; Gupta 1968; Reddy and D'Cruz, 1969 a,b; Gupta *et al.*, 1969)	Apospory Panicum and Hieracium types parthenogenesis pseudogamous	Facultative 4-nucleate
D. aristatum (Celarier and Harlan, 1957)	Apospory parthenogensis pseudogamous	Facultative, 4-nucleate
D. caricosum (Celarier and Harlan, 1957)	Apospory parthenogensis pseudogamous	Facultative, 4-nucleate
D. intermedium (2n = 40) (Saran and de Wet, 1970)	Apospory parthenogensis pseudogamous	parthenogensis 4-nucleate
D. nodosum (Carman, 1995)	Apospory	—
D. papillosum (Calarier and Harlan, 1957)	Apospory pseudogamous	Facultative, 4-nucleate

(contd.)

(Table IV contd.)

Species investigated and author	Type of Apomixis	Mechanism of Apomixis
Echinochloa frumentacea (Narayanaswami, 1955b)	Apospory	Facultative
E. stagnina (2n = 36, 54, 72, 108, 126) (Muniyamma, 1978)	Apospory	Facultative 4-nucleate
Elymus rectisetus (Hair, 1956; Crane and Carman, 1987)	Diplospory Taraxacum type	Both facultative and obligate forms occur
Eragrostiella bifaria (Venkateswarlu & Devi, 1964; Sonpipare, 1984)	Apospory	Facultative
Eragrpstis barbinodis (Voigt and Bashaw, 1972, 1976; Burson and Voigt, 1996)	Diplospory Pseudogamy	—
Eragrostis bicolor (Brown and Emery, 1958; Streetman, 1963)	Diplospory	Obligate, 4-nucleate
Eragrostis caesia (Voigt and Bashaw, 1972, 1976; Bursoo and Voigt, 1996)	Diplospory Pseudogamy	—
E. chloromelas (Brown and Emery, 1958; Streetman, 1963; Voigt and Bashaw, 1972, 1976)	Apospory	4-nucleate
E. curvula (2n = 20, 40) (Streetman, 1963; Brown and Emery, 1958b; Voigt and Bashaw, 1953; Brix, 1974; Burson and Voigt, 1996)	Apospory or rarely Diplospory	Facultative 4-nucleate
E. heteromera (Brown and Emery, 1958)	Apospory	4-nucleate
E. intermedia (Brown and Emery, 1958; Streetman, 1963)	Apospory	4-nucleate
Eragrostis planiculmis (Voigt and Bashaw, 1972, 1976; Burson and Voigt, 1976)	Diplospory Pseudogamy	—
E. plano (Brown and Emery, 1958; Streetman, 1963)	Apospory	4-nucleate

Species investigated and author	Type of Apomixis	Mechanism of Apomixis
Eragrostis rigidior (Voigt and Bashaw, 1972, 1976; Burson and Voigt, 1976).	Diplospory Diplospory	—
E. superba (Brown and Emery, 1958; Streetman, 1963)	Apospory	4-nucleate
Eremopogon foveolatus (Bhanwra and Choda, 1981; Satyamurty and Seshavatharam, 1983)	Apospory	Facultative 4-nucleate
Eriochloa borumensis (Brown and Emery, 1958)	Apospory	Facultative, 4-nucleate
Eriochloa procera (Shanthamma and Narayan, 1976–77, Seshavathram, 1983)	Apospory Panicum type	Facultative
E. sericea (Brown and Emery, 1958	Facultative	4-nucleate
Eustachys paspaloides (2n=40) (Strydom and Spies, 1994)	Apospory Hieracium type	Facultative
Festuca drymeja (Shishkinskaya & Borodko, 1987)	Apomictic	—
Fingerhuthia africana (Brown and Emery, 1958; Kellog, 1990)	Apospory	Facultative 4-nucleate
Harpochloa fallx (2n-40, 60) (Strydom and Spies, 1994)	Apospory Hieracium type	Facultative
Heteropogon contortus (Gupta, 1968; Tothil, 1968; Emery and Brown, 1958)	Apospory parthenogenesis pseudogamy	Facultative, 4-nucleate
Hierochloe alpina (2n = 57, 66) (Weimarck, 1970)	Diplospory	Facultative, 4-nucleate
H. australis (Weimarck, 1967a)	Apospory Parthenogenesis	Facultative, 8-nucleate
H. monticola (2n = 63) (Weimarck, 1967b, 1970)	Apospory Parthenogenesis	Apospory, 8-nucleate
H. odorata (Norstog, 1963; Weimarck, 1967a)	Apospory, pseudogamous	4-nucleate Obligate
Hierochloe wendelboi (2n = 28) (Weimarck, 1977)	Apospory	8-nucleate
Hilaria belangeri (Brown and Coe, 1951)	Apospory	8-nucleate
H. mutica (Brown and Coe, 1951)	Apospory	Facultative
Hyparrhenia hirta (Brown and Emery, 1958; McWillam *et al.*, 1970)	Apospory	4-nucleate

(contd.)

(Table IV contd.)

Species investigated and author	Type of Apomixis	Mechanism of Apomixis
H. rufa (Brown and Emery, 1958)	Apospory	4-nucleate
Ischaemum indicum (Seshavatharam, 1983)	Apospory Panicum type Pseudogamy	Facultative 4-nucleate
I. laxum (Seshavatharam, 1983)	Apospory Hieracium type	Pseudogamy Facultative
Lamprothyrsus peruvianus (Connor and Dawson, 1993)	Apospory Hieracium type Pseudogamy	Facultative
Merxmuellera arundinacea (Verboom *et al.*, 1994)	Apomictic	—
M. rufa (Verboom *et al.*, 1994)	Apomictic	—
Nardus stricta (Rychlewski, 1961)	Diploid apomict	Facultative
Panicum antidotale (Warmke, 1954; Shamakumari, 1960)	Apospory, Hieracium type	Facultative, 8-nucleate
P. deustum (Brown and Emery, 1958)	Apospory	4-nucleate
P. maximum (2n = 32) (Warmke 1954; Brown and Emery, 1958; Pernes *et al.*, 1975; Savidan, 1975, 1978; Savidan *et al.*, 1979)	Apospory Pseudogamous	Facultative, 4 and 8-nucleate
P. notatum (Febulaus, 1992)	Aposporous Parthenogenesis	Obligate, 8-nucleate
P. notatum (Brown and Emery, 1958)	Apospory	4-nucleate
Panicum obtusum (Carman, 1995)	Apospory	—
P. virgatum (Brown and Emery, 1958)	Apospory	4-nucleate
Paspalum alcalinum (2n = 50*) (Burson, 1997) * Diploids (2n = 20) are sexual while tetraploids; (2n = 40) are facultative apomicts	Apospory	Obligate apomic
Paspalum almum (2n = 24) (Burson, 1975)	4-nucleate Pseudogamy	Obligate
Paspalum arundinellum (Bashaw *et al.*, 1970)	Apospory	—
Paspalum atratum (2n+40) (Quarin *et al.*, 1997)	Apospory Parthenogensis	Obligate

Species investigated and author	Type of Apomixis	Mechanism of Apomixis
Paspalum brunneum (2n = 40) (Norrmann *et al.*, 1989; Burson and Bennett, 1971)	Apospory Pseudogamous	4-nucleate, Obligate
Paspalum commersonii (2n = 20, 40, 60) (Chao, 1974)	Diplospory Parthenogenesis	Parthenogenesis
*Paspalum compressifolium** (2n = 40, 60) (Quarin *et al.*, 1996) *Diploids (2n=20) are sexual	Apospory	Obligate
P. conjugatum (Chao, 1980)	Parthenogenesis	4-nucleate
P. coryphaeum (2n = 40) (Quarin & Urbani, 1990; Burson and Bennett, 1971)	Pseudogamous	Obligate
P. cromyorhizon (2n = 40) (Burson and Bennett, 1971a)	Apospory Apospory	4-nucleate, Obligate
Paspalum densum (Caponio & Quarin, 1993)	Apospory	Facultative only in 10% of ovules
P. dilatatum (2n = 6X = 60)* (Bennet, 1944, Smith, 1948, Bashaw and Holt, 1958, Brown and Emery, 1958; Burson *et al.*, 1991) *Tetraploids (2n = 4X = 40) of *Paspalum dilatatum* are sexually reproducing	Apospory	Obligate, 4-nucleate
Paspalum dedeccae (2n = 4X = 40) (Quarin and Burson, 1991)	Apospory Pseudogamous	4-nucleate Obligate
Paspalum densum (Caponio & Quarin, 1993)	Apospory	Facultative
Paspalum denticulatum (2n = 4X = 40) (Quarin and Burson, 1991)	"	"
P. distichum (2n = 60) (Choda and Bhanwra, 1977; Bhanwra, 1988; Burson and Bennet, 1971a)	Apospory	Obligate 8-nucleate
Paspalum durifolium (2n = 6X = 60) (Burson, 1985)	Apospory Pseudogamous	4-nucleate Obligate
P. exaltatum (2n = 40) (Burson and Bennett, 1971a)	8-nucleate Pseudogamous	Obligate
Paspalum falcatum (2n = 40) (Burson, 1997)	Apospory	Facultative

(contd.)

(Table IV contd.)

Species investigated and author	Type of Apomixis	Mechanism of Apomixis
Paspalum guaraniticum (2n = 80) (Burson and Bennet, 1970b)	Apospory	Facultative
P. guenoarum (2n = 36) (Pritchard, 1970; Burson and Bennett, 1971)	Apospory Pseudogamous	Obligate, 8-nucleate
P. hartwegianum (Carman, 1995)	Apospory	—
P. haumanii (2n = 40)* (Norrmann, *et al.*, 1989; Burson, 1975; Burson and Bennett, 1971) * Diploids are sexual	Apospory Pseudogamous	4-nucleate, Obligate
P. hexastachyum (2n = 40) (Quarin & Hanna, 1980)	Apospory	—
Paspalum hydrophilum (Norrmann, 1981)	Apospory	—
Paspalum jesuiticum (Burson, 1975)	Apospory	—
P. intermedium (2n = 40) (Norrmann *et. al.*, 1989)	Apospory	Facultative
Paspalum lividum (2n = 40) (Burson and Bennet, 1971a)	Diplosporic Parthenogensis	Obligate, 8-nucleate
Paspalum jesuiticum (Burson, 1975)	Apospory	—
P. longifolium (2n = 40, 50, 60) (Chao, 1974)	Apospory Pseudogamous	Facultative
P. maculosum (2n = 40) (Norrmann *et al.*, 1989)	Apospory Pseudogamous	Pseudogamous Facultative
P. malacophyllum (Brown and Emery, 1958; Burson and Hussey, 1998)	Apospory Pseudogamous	4-nucleate Obligate
P. mandiocanum (2n = 60) (Carman, 1995; Burson and Bennett, 1971a)	Apospory	4-nucleate
Paspalum minus (Bonilla & Quarin, 1997)	Diplospory Apospory	Parthenogenesis
P. nicorae (2n = 40) (Burson and Bennett, 1970a)	Pseudogamy	Obligate
Paspalum notatum (Burton, 1948)	Apomictic Pseudogamy	Obligate
P. orbiculare (2n = 40, 50, 60) (Chao, 1964)	Diplosporic	Obligate
P. paspaloides (Carman, 1995)	Diplosporic	Obligate

Species investigated and author	Type of Apomixis	Mechanism of Apomixis
Paspalum pauciflorum (Burson, 1997)	Apospory	Facultative
P. plicatulum (2n = 36) (Pritchard, 1970; Burson and Bennett, 1971b)	Apospory Pseudogamous	4-nucleate, Obligate
Paspalum polyphyllum (Burson, 1997)	Apospory	Facultative
P. proliforum (2n = 60) (Quarin *et al.* 1982; Burson, 1975)	Apospory Pseudogamous	4-nuclcate, Obligate
P. quadrifarium (2n = 40) (Norrmann *et al.*, 1989)	Apospory Pseudogamous	4-nucleate, Obligate
P. quaraniticum (2n = 40, 50) (Burson and Bennett, 1970)	Apospory	4-nucleate
Paspalum ramboi (Quarin'and Burson, 199)	Apospory Pseudogamous	4-nucleate Obligate
Paspalum rojasii (Burson and Bennett, 1971b)	Apospory, Pseudogamous	4-nucleate, 4-nucleate
P. rufum (2n = 40) (Norrmann *et al.*, 1989; Burson, 1975)	Apospory Pseudogamous	4-nucleate, Obligate
P. scrobiculatum (Carman, 1995)	Apospory	—
P. secans (Emery, 1957, Snyder, 1957)	Apospory Obligate,	Apospory, 8-nucleate
Paspalum simplex (Pupili *et al.*, 1997)	Apospory	—
P. thunbergii (2n = 20, 40, 60) (Chung, 1974)	Diplospory, Parthenogenesis	8-nucleate
Paspalum unispicum (2n = 40) (Burson, 1997)	Apospory	Facultative
Pennisetum clandestinium (2n=36), (Narayan, 1958, Emery, 1957)	—	Facultative, 4-nucleate
P. dubium (2n=66) (Gildenhuys and Brix, 1959)	Apospory Pseudogamous	Facultative
Pennisetum flaccidum (2n = 36) (Bashaw and Hussey, 1992; Chatterji and Timothy, 1969)	Apospory	4-nucleate
P. hohenackeri (2n=18) (Kurup, 1983, Seshavatharam, 1983)	Panicum type Pseudogamous	Obligate
P. massaicum (Saran and Mishra, 1986; D' Cruz and Reddy, 1968)	Apospory	Facultative, 4-nucleate

(contd.)

(Table IV contd.)

Species investigated and author	Type of Apomixis	Mechanism of Apomixis
P. mezianum (2n=32) (Shanthamma, 1974, Shanthamma and Narayan, 1976; Bashaw and Hussay, 1992)	Apospory	Obligate
P. orientale (2n=27) (Narayan, 1951, Choda *et al.*, 1981, Emery 1957; Bhanwra, 1988)	Apospory parthenogenesis pseudogamous	4-nucleate; Obligate or Facultative rarely 5-nucleate
Pennisetum pedicellatum (Chatterji and Pillai, 1970; Shobha, 1988)	Apospory Parthenogenesis	Facultative, 4-nucleate
P. polystachyon (2n=63) (Muniyamma, 1973; Birari, 1981; Saran & Mishra, 1986)	Apospory Panicum type Pseudogamous	Faculative, 4-nucleate
P. purpureum (2n=28) (Brown and Emery, 1958)	Apospory	4-nucleate, Facultative
P. ramosum (2n=20) (Narayan, 1962)	Diploid apomict	Facultative
P. reuppelli (Narayan, 1951; Reddy, 1977)	Apospory Parthenogenesis	Facultative, 8-nucleate and 4-nucleate
P. setaceum (2n = 27, 54) (Narayan, 1951; Emery, 1957; Simpson and Bashaw, 1969)	Apospory Parthenogenesis	Obligate, 4-nucleate
Pennistum squamulatum (2n = 54 + 2) (Sindhe, 1976)	Apospory Parthenogensis	Facultative, 4-nucleate
P. villosum (2n=45) (Naryan, 1951; Fischer *et al.*, 1954; Snyder *et al.*, 1955; Gildenhuys and Brix, 1959)	Apospory Parthenogenesis	Obligate, 8- and 4-nucleate
Pentaschistis eriostoma (Spies *et al.*, 1999)	Apospory Panicum type	Facultative 4-nucleate
Poa alpina (Hakansson, 1943; 1944, Nygren, 1967)	Diplospory Parthenogenesis Pseudogamous	Facultative, 8-nucleate
P. ampla (Kellogg, 1987)	Diploid apomict Apospory	Obligate
P. arctica (Engelbert, 1941, Nygren, 1950b)	Apospory, Parthenogenesis	Facultative
Poa arida (Carman, 1995)	Apospory	—
P. canbyi (Kellogg, 1987)	Apospory Parthenogenesis	Facultative

Species investigated and author	Type of Apomixis	Mechanism of Apomixis
P. compressa (Armstrong, 1937)	Apospory	—
P. curtifolia (2n = 42) (Kellogg, 1987)	Apospory	Facultative
P. glauca (Nygren, 1950a, b; Akerberg, 1941)	Apospory	Facultative
P. gracillima (Kellogg, 1987)	Apospory	Facultative
P. granitica (Skalınska, 1959)	Apospory	Facultative
Poa herjedalica (*alpina x pratensis*) (Carman, 1995)	Apospory	—
P. incurva (Kellogg, 1987)	Apospory	Facultative
P. juncifolia (Kellogg, 1987)	Apospory	Obligate
P. nemoralis (Nygren, 1950a,b)	Aposproy	—
P. nervosa (Grun, 1955)	Apospory	Obligate, 8-nucleate
P. nevadensis (Kellogg, 1987)	Apospory	Facultative
P. palustris (Kiellander, 1937)	Apospory	—
P. pratensis (Tinney, 1940, Kiellander, 1941, Akerberg, 1942; Nielson, 1946; Grazi *et al.*, 1961)	Diplospory Pseudogamous	Facultative, 8-nucleate
P. sandbergii (Kellogg, 1987)	Diplospory Pseudogamous	Facultative
P. scabrella (Kellogg, 1987)	Diplospory Pseudogamous	Facultative
P. secunda (Kellogg, 1987)	Apospory Pseudogamous	Facultative
P. serotiana (Kiellander, 1935)	Apospory	—
Rendlia altera (Strydom and Spies, 1994) (2n=40)	Apospory Hieracium type	Obligate
Rhynchelytrum villosum (2n=36) (Shanthamma and Narayan, 1976-77)	Aposporus	Facultative
Saccharam officinarum (Narayanaswami, 1940)	Diplospory	Facultative, 8-nucleate
S. spontaneum (Narayanaswami, 1940)	Diplosporous	Facultative
Schizachyrium sanguineum (Carman and Hatch, 1982)	Apospory	—
Schima ischaemoides (Seshavatharam, 1983)	Apospory Panicum type Pseudogamous	Facultative, 4-nucleate
Setaria leucopyla (Emery, 1957) rarely	Apospory	Facultative, 4-nucleate 8-nucleate
S. macrostachya (Emery, 1957; Reddy, 1977)	Apospory Parthenogenesis Pseudogamous	8-nucleate

(contd.)

(Table IV contd.)

Species investigated and author	Type of Apomixis	Mechanism of Apomixis
S. villosissima (Emery, 1957)	Apospory	Facultative, 4-nucleate 8-nucleate
Sorghum bicolor (Rao and Narayan, 1968; Ross and Wilson, 1969; Marshall and Downes, De Wet, 1978; Murthy *et al.* 1979; Hanna *et al.*, 1970; Reddy *et al.*, 1979; Wu *et al.*, 1994)	Apospory, Hieracium type Pseudogamous Diplospory	Facultative, 8-nucleate
S. halepense (Seshavatharam, 1983)	Apospory Panicum type Pseudogamous	Facultative, 4-nucleate
Themeda arundinacea (Birari, 1981)	Apospory	—
T. australis (2n = 20, 40) (Evans and Knox, 1969)	Apospory Parthenogenetic	4-nucleate
T. dacruzi (Birari, 1981)	Apospory	—
T. longispatha (Zagorcheva, 1989)	Apospory	—
T. hexandra (Brown and Emery, 1956)	Apospory Panicum type Pseudogamous	—
T. quadrivalvis (Brown and Emery, 1958)	Apospory	4-nucleate
*T. triandra** (Brown and Emery, 1958; Liebenberg, 1990) *Diploids are sexual	Apospory Parthenogenesis Pseudogamous	4-nucleate
Trachys muricata (Seshavatharam, 1983)	Apospory Panicum type Pseudogamous	Facultative
Tribolium echinatum (Visser and Spies, 1994)	Diplospory Hieracium type	Facultative
T. uniolae (Visser and Spies, 1994)	Diplospory Hieracium type	Facultative
T. utriculosum (Visser and Spies, 1994)	Diplospory Hieracium type	Facultative
Tricholaena monachne (Carman, 1995)	Apospory	—
Tripsacum bravum (Leblanc *et al.*, 1995)	Diplospory	—
T. dactyloides (2n = 72)* (Farquharson, 1955; Chen, 1994; Burson *et al.*, 1990; Kindiger and Dewald, 1994) * Diploids of *Tripsacum dactyloides* (2n = 36) are sexual	Diplospory Hieracium type Pseudogamous	Facultative 4-nucleate

Species investigated and author	Type of Apomixis	Mechanism of Apomixis
T. intermedium (Leblanc *et al.*, 1995)	Diplospory	—
T. lanceolatum (Leblanc *et al.*, 1995)	Diplospory	—
T. maizar (Leblanc *et al.*, 1995)	Diplospory	—
T. pilosum (Leblanc *et al.*, 1995)	Diplospory	—
T. zopilotense (Leblanc *et al.*, 1995; Leblanc & Savidan, 1994)	Diplospory	—
Urochloa bolbodes (Brown and Emery, 1958)	Apospory	4-nucleate
U. mosambicensis (2n = 36) (Brown and Emery, 1958; Pritchard, 1970)	Apospory	4-nucleate
U. panicoides (Khan, 1990)	Apospory	Facultative
U. pullulans (Brown and Emery, 1958)	Apospory	4-nucleate
U. trichopus (Brown and Emery, 1958)	Apospory	4-nucleate
Zea mexicana (Tantravahi, 1965)	—	Facultative

(u hybrids); (3) the reduced egg cells by parthenogensis, give rise to (poly) haploids and (4) the reduced egg cells, through fertilization, produce variable sexual offsprings (R-hybrids).

Organisation of the Aposporic Embryo Sacs

The organization of the mature aposporic embryo sacs in Poaceae can be studied under two types—the Pennisetum type (Panicum type) and the Hieracium type.

i) Pennisetum Type (=Panicum type)

In this type one or more nucellar cells develop into aposporic embryo sacs. After vacuolation the nucleus of the aposporous initial undergoes two divisions. All the four nuclei resulting from this division remain at the micropylar region only. Then wall formation takes place giving rise to two synergids, 1 egg cell and 1 polar nucleus. This type hitherto was called as Panicum type. But it was Narayan (1951) who reported it for the first time in *Pennisetum villosum, P. setaceum* and *P. orientale*. As this work was in the Ph.D. thesis, the credit was given to Warmke (1954) who reported this type in *Panicum maximum* and consequently reported as Panicum type. Battaglia (1992) pointed out this and said that this type should be named as the Pennisetum type and credit should be given to Narayan. Pennisetum type is known only in Panicoideae of family Poaceae, but it is very widespread. It occurs in *Bothriochloa* spp. (Celarier and Harlan, 1957), *Capillepedium* spp. (Celarier and Harlan, 1957; Snyder *et al.*, 1955; Shanthamma, 1982), *Dichanthium* spp (Celarier & Harlan, 1957), *Cortaderia jubata* (Philipson, 1978), *Eragrotis* spp. (Streetman, 1963; Brown & Emery, 1958), *Anthephora pubescens* (Brown & Emery, 1958), *Brachiaria brizantha, B. serrata* (Brown & Emeryy, 1958, *Bouteloua curtipendula* (Mohammed & Gould, 1966), *Cenchrus, Chloris, Digitaria, Eremopogon foveolatus* (Satyamurthy & Seshavatharam, 1983), *Eriochloa* spp. (Brown & Emergy, 1958), *Heteropogon contortus* (Gupta, 1968), *Hierochloe* spp. (Weimarck, 1970), *Hyparrhenia, Panicum* spp. (Warmke, 1954; Brown & Emery, 1958, Febulaus, 1992), *Paspalum* spp. (Brown & Emergy, 1958; Chao, 1974), *Pennisetum* spp. (Narayan, 1951; Brown & Emergy, 1958; Shanthamma, 1974), *Sorghum, Themeda, Urochloa* (Brown & Emery, 1958) etc.

Reddy (1977) reported that most of the apomicts of Panicoideae, with few exceptions are aposporous and produce mostly 4-nucleate embryo sacs. Brown and Emery (1958) stated that the unreduced embryo sacs in the Panicoideae are 4-nucleate and conversely that all the 4-nucleate embryo sacs in the Panicoideae are unreduced. 4-Nucleate unreduced embryo sacs have been reported in *Pennisetum reupelli* and *P. villosum* (Narayan, 1951), *P. ciliare* and *Cenchrus setigerus* (Fisher *et al.*, 1954), *C. ciliaris* (Fig. 19A,B,D-F) (Snyder *et al.*, 1955; Febulaus and Pullaiah, 1995), *Pennisetum orientale, P. villosum, P. clandestinum, P. setaceum* (Emery, 1957), *Setaria leucopyla, S. villosissima* (Emery, 1957). Brown and Emery (1958) reported the presence of 4-nucleate aposporic embryo sacs in several species of the tribe paniceae.

The *Pennisetum* generic complex as a whole consists of diploid and polyploid apomicts which vary from almost facultative apomicts (*P. purpureum* 2n = 28, 4 nucleate type, Brown and Emery, 1958; *P. dubium* 2n = 66, Gildenhuys and Brix, 1959; *P. hohenackeri* 2n = 18, Kurup, 1973; *P. polystachyum* 2n = 63, Muniyamma 1973; *P. clandestinum* 2n = 46, *P. ramosum* 2n = 20, *P. orientale* 2n = 27, Narayan, 1951; *P. ciliare* 2n = 36, 40, 54, Snyder *et al.* 1955; *P. squamulatum* 2n = 54, Sindhe 1976) to obligate apomicts (*P. villosum* 2n = 45, *P. setaceum* 2n = 27, Narayan, 1951 and *P. mezianum* 2n = 32, Shanthamma, 1973). The only investigated diploid sexual species is *P. typhoides* (2n = 18, Narayanaswamy, 1953) while *P. macrostachyum* is a tetraploid (2n = 68) and sexually reproducing species (Shanthamma, 1979). Similarly *Pennisetum latifolium* (2n = 36) is sexually reproducing (Narayan, 1951).

In *Capillipedium huegelli* (Choda and Bhanwra, 1980) the 4-nucelate aposporous embryo sacs are of two types. The mature aposporous embryo sacs may show an egg, two synergids and single polar nucleus or one egg, one synergid and two polar nuclei. Such organization of embryo sac is also reported in *Eremopogon foveolatus* (Satyamurthy and Seshavatharam, 1983). 4-nucleate aposporous embryo sacs are also recorded in *Pennisetum orientale, Bothriochloa*

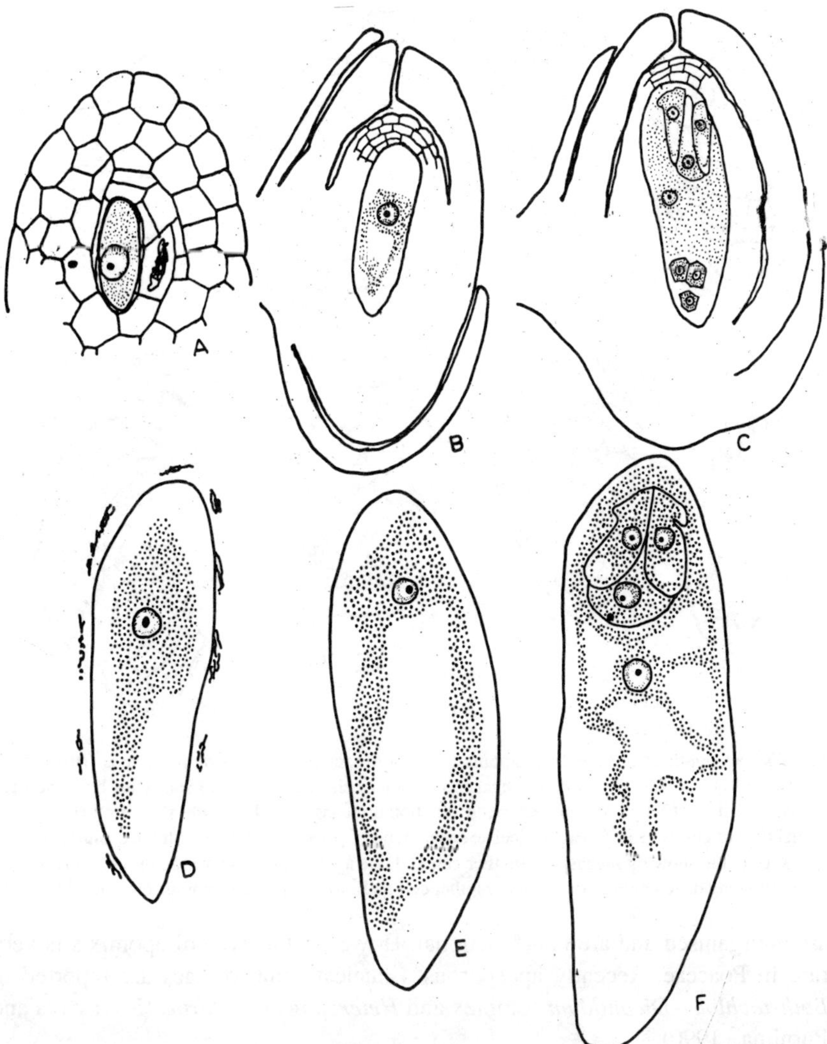

Fig. 19. A–F. Aposporic embryo sac development in *Cenchrus ciliaris* (Febulaus & Pullaiah, 1995).

odorata, Capillipedium huegelli, C. parviflorum, Eremopogon foveolatus (Bhanwra, 1988), *Pennisetum polystachyon* (Sunil Saran and Mishra, 1986).

In *Arundinella mesophylla* the organization of the aposporous embryo sacs does not confirm to any of the known types. The nuclei in the mature embryo sac show either 1 + 2 or 2 + 5 arrangement or even they may aggregate and remain in the centre and are non functional (Basappa and Muniyamma, 1981). In *Echinochloa stagnina* (Muniyamma, 1978) also the aposporous embryo sacs

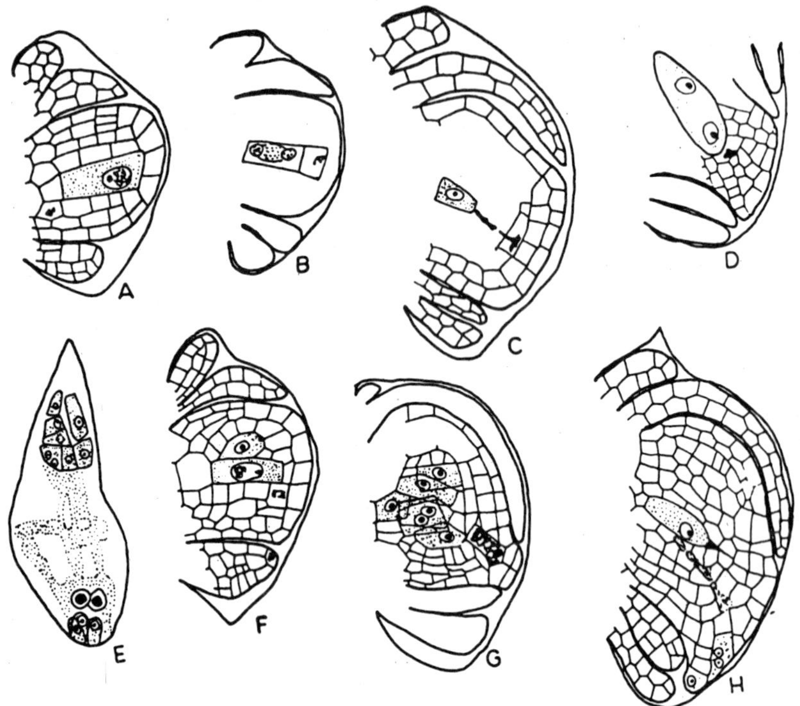

Fig. 20. Development of sexual and apomictic embryo sacs in *Eremopogon foveolatus* (Bhanwra & Choda, 1981). A. L.s. of ovule at megaspore mother cell stage. B. Dyad stag with both the cells dividing further. C. L.s. of ovule showing functional megaspore. D. Ovule showing two nucleate sexual type of embryo sac. E. Eight nucelate sexual type of embryo sac with proliferated antipodal cells. F. L.s. of ovule showing megaspore mother cell and one aposporous embryo sac initial. G.H. L.s. of ovule showing degenerating megaspore mother cell and formation of aposporous embryo sac initials.

are unorganised and are non-functional. However, this type of apomixis is very rare in Poaceae. Recently aposporous 4-nucleate embryo sacs are reported in *Bothriochloa—Dicanthium* complex and *Heteropogon contortus* (Srivastava and Purnima, 1990).

ii) Hieracium Type

In this type one or more somatic cells begin to enlarge and develop directly into the initials of aposporous embryo sacs (Fig. 21E). The nuclear divisions take place and results in the 8-nucleate aposporous embryo sac having an egg apparatus, 2 polar nuclei and three antipodals. 8-Nucleate aposporous embryo sacs have been reported in *Poa compressa* (Armstrong, 1937), *P. pratensis* (Akerberg, 1939; 1942; 1943; Tinney, 1940; Muntzing, 1940; Nygren, 1950b), *P. arctica* (Nygren, 1950b), *Paspalum secans* (Snyder, 1957), *Pennisetum*

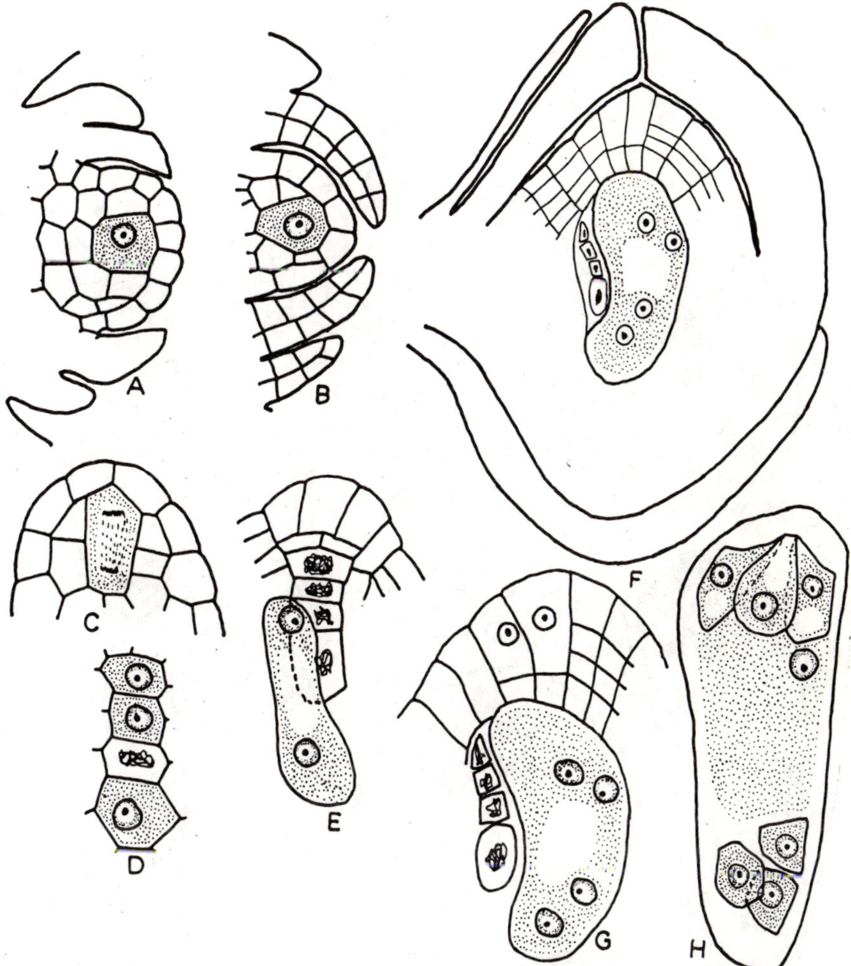

Fig. 21. Aposporic embryo sac development in *Panicum notatum* (Febulaus, 1992).

clandestinum (Narayan, 1955), *Panicum antidotale* (Warmke, 1954; Shamakumari, 1960), *Hierochloe* (Fig. 22) (Weimarck, 1970). *Paspalum commersonii* (Chao, 1974), *P. paspaloides* (Srivastava and Purnima, 1990), *Panicum notatum* (Fig. 21A–H) (Febulaus and Pullaiah, 1991). In the *Sorghum* variety R473 (N.G.P. Rao *et al.* 1978), the female archesporial cell directly develops into megaspore mother cell which degenerates. Therefore, sexual embryo sacs are very rare. The embryo sac develops from a nucellar cell and Hieracium type of apospory occurs. The pollen tubes grow to a very short length in the stigma and do not reach up to the ovary. Due to the stimulus of pollimation, endosperm and embryo develop autonomously. The apomixis in variety R473

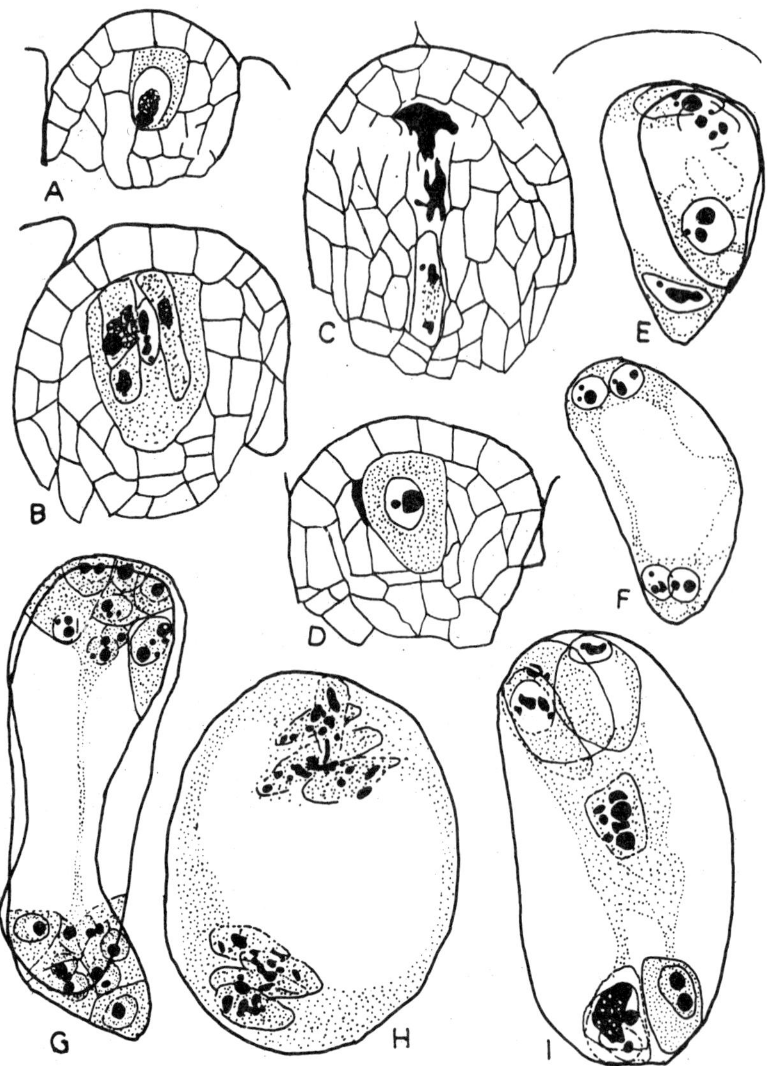

Fig. 22. Apomictic embryo sac development in *Hierochloe* (Weimarck, 1970)

is associated with self-incompatibility. Like the two parents the Fl progeny exhibits somatic apospory associated with self-incompatibility. In *Panicum maximum* (Warmke, 1954), *Pennisetum ciliare* (Snyder *et al.,* 1955), *Dicanthium annulatum* (Reddy and D'Cruz, 1969), *Sorghum bicolor* (Murthy and Rao, 1972), *Bothriochloa ischaemum* (Moskova and Yakovlev, 1974), the 8-nucleate aposporous embryo sacs of Hieracium type were found to occur along with the 4-nucleate aposporous embryo sacs of Panicum type. The antipodals of the aposporous 8-nucleate embryo sacs persist and proliferate as in the case of sexual embryo sacs (Seshavatharam, 1977).

Organization of aposporous embryo sacs with three nuclei, i.e., an egg and two polars has been reported in *Paspalum dilatatum* (Bashaw and Holt, 1958), *Pennisetum ciliare* (Bashaw, 1962), *Bouteloua curtipendula* var. *caespitosa* (Mohamed and Gould, 1966), *Sorghum bicolor* (Bashaw, 1970). However these observations need confirmation.

2. Diplospory

Development of embryo sac from generative cells without meiosis or with irregular meiosis is known as Diplospory. This includes the terms Generative apospory (development of unreduced embryo sac from generative cell without meiosis) and Aneuspory (development of unreduced embryo sac from generative cell with irregular meiosis) of Battaglia (1963).

Antennaria Type

In this type the megaspore mother cell does not enter meiosis, but after a long interphase functions as unreduced (diploid) megaspore. It undergoes three mitotic divisions resulting in an 8-nucleate embryo sac with 3-celled egg apparatus, 3 antipodals and 2 polar nuclei. In the family Poaceae Antennaria type has been reported in *Poa alpina* (Hakansson, 1943), *Poa nervosa* (Grun, 1955), *Calamogrostis chalybaea* (Stenar, 1932; Nygren, 1946), *C. lapponica* (Nygren, 1946), *C. purpurascens* (Nygren, 1954), *C. crassiglumis* (Nygren, 1954), *C. stricta* (Greene, 1984), *C. stricta* sub sp. *inexpansa* (Nygren, 1954; Greene, 1984), *C. canadensis* (Nygren, 1954; Jorgensen *et al.*, 1958), *Bouteloua* (Snyder and Harlan, 1953), some biotypes of *Paspalum commersonii* and *P. longifolium* (Chao, 1974), *Nardus stricta* (Rhychlewski, 1961), *Tripsacum dactyloides* (Farquharson, 1955; Leblanc *et al.*, 1995), *Poa pratensis* (Kiellander, 1941), *Eragrostis chloromelas, E. curvula, E. lehmaniana* (Streetman, 1963), *Hierochloe alpina* (Weimarck, 1970) and in *Cortaderia jubata* (Costas-Lippmann, 1979).

Taraxacum Type

In this type the megaspore mother cell undergoes irregular meiosis I resulting in restitution nucleus. The second meiotic division starts with unreduced chromosome numbers and results in a unreduced dyad instead of a reduced tetrad. The chalazal dyad cell is functional and undergoes three mitotic divisions resulting in an 8-nucelate embryo sac. This type has been recorded in *Agropyron scabrum* (Hair, 1956), certain species of *Paspalum* (Chao, 1974, 1980), apomictic forms of *Elymus reetisetus* (Crane and Carman, 1987) and *Tripsacum dactyloides* (Leblanc *et al.*, 1995).

From the above observations it is well established that the organised aposporous embryo sacs in Poaceae are either 4-nucleate or eight nucleate (Stebbins, 1950; Brown and Emery, 1958; Battaglia, 1963; Nygren, 1967). Unreduced 4-nucleate embryo sacs are reported in some species of the tribes Chlorideae and Eragrosteae (Brown and Emery, 1958), while they are 8-nucleate in the tribes Agrostideae, Aveneae, Festuceae and Triticeae (Nygren, 1946; 1949; 1951; Hakanseen, 1943; 1944; 1948; Kiellander, 1937; Weimarck, 1947a,b; 1970). In Paniceae 14 species are completely aposporous and appear to be both sexual and aposporous. In Andropogoneae apomixis has been reported in 17 species belonging to 6 genera viz. *Bothriochloa, Capillipedium* and *Dicanthium*. However other genera of this tribe where apomixis has been reported include *Apluda, Heteropogon, Hyparhenia, Saccharum, Sorghum* and *Themeda* (Brown and Emery, 1958; Murty, 1973 and Narayanaswami, 1940). No apomictic species have been reported from the genera *Andropogon, Eremochloa, Erianthus, Iseilema, Rottboellia,* and *Sorghastrum* (Bhanwra, 1988). The other details of Apomixis in Poaceae have been shown in the Table IV.

Development of the Endosperm and Embryo

In the apomictic embryo sacs the development of endosperm is associated with pollination and the fusion of the sperm nucleus with one of the polar nuclei in most of the reported cases of obligate and facultative apomicts (Seshavatharam, 1977). When pollination does not occur, the endosperm develops from the polar nuclei without fertilization. This mode is referred to as "autonomous" and the egg develops parthenogenetically, e.g. *Cortaderia jubata*. Autonomous development of endosperm was reported in some diplosporous species of *Calamogrostis* (cf. Nygren, 1967). Murty and Rao (1972) also reported the autonomous development of endosperm in the self-incompatable strains of r 473 of *Sorghum bicolor.*

Development of Embryo

The development of unreduced egg may follow semigamous, pseudogamous, parthenogenic or autonomous manner.

1. Semigamy

Rao and Narayana (1980) reported Semigamy for the first time in the family Poaceae. According to them in *Coix aquatica* (n = 5) independent divisions of egg and sperm nuclei, accompanied by cell wall formation, led to the formation of maternal and paternal tissues, that after having been separated from one another continued growth each in its own way and resulted in a forked seedlings. The observation of forking of the seedling stem with the constituents of forking differing genetically and cytologically and resembling the male and female parents led to the identification of Semigamy.

Incidentally, this report of semigamy in *Coix aquatica* is the fourteenth case among Angiosperms, the sixth in Monocotyledons, and the first in grass family. Solntzeva (1974) who prefers to call the phenomenon hemigamy on linguistic grounds, pointed out that in semigamy the sexual nuclei preserve their autonomous development longer and further development may lead to the fusion of male and female nuclei implying that it is only a prolonged process of fertilization, since in normal fertilization male and female nuclei fuse the moment the sperm enters the egg. However, in *Coix aquatica* (Rao and Narayana, 1980) in the study of Semigamy, the fusion of male and female nuclei (after having had some autonomous development) was not apparent; on the contrary, the paternal and maternal tissues derived from semigamy, after having been together for sometime, diverged into distinct branches as seen in pseudo-twins.

2. Pseudogamy

Here pollination takes place, but male nucleus degenerates inside the egg. Embryo develops from the egg. Pseudogamy has been reported in *Hierochloe* (Weimark, 1967), *Panicum maximum* (Warmke, 1954), *P. ciliare* (Snyder *et al.*, 1955), *P. dubium* (Gildenhuys and Brix, 1959).

In majority of cases the development of unreduced egg follows pseudogamy (Battaglia, 1963). It has been reported in *Eragrostis curvula, E. intermedia, E. lehmanniana, E. plano, E. superba* (Streetman, 1963), *Capillipedium huegelii* (Choda and Bhanwra, 1980), *Hierochloe odorata* (Norstog, 1963), *Bouteloua curtipendula* (Harlan, 1949), *Paspalum secans* (Snyder, 1957), *Pennisetum ciliare* (Snyder *et al.*, 1955) and *Eremopogon foveolatus* (Bhanwra and Choda, 1981).

Reddy and D'Cruz (1969n) described Pseudogamous development in *Dichanthium annulatum*. In four nucleate embryo sacs the polar nucleus is always fertilized by the male gamete and the other male gamete either enters the egg cell or remains outside (Pseudogamy). In eight-nucleate unreduced embryo sacs both the polar nuclei get independently fertilized by the male gamete. Frequently syngamy is eliminated due to (1) failure of the sperm nucleus to enter the egg, (2) degeneration of the male nucleus inside the egg cell, (3) non-availability of the male gamete for fertilizing the egg cell in eight nucleate embryo sac where both the polar nuclei are independently fertilized, and (4) development of nucellar embryos. Reports of Pseudogamy in Poaceae are not rare (see Choda and Bhanwra, 1977, Grun, 1955; Nygren, 1954; Rychlewski, 1966, Shantamma, 1982).

3. Parthenogenesis

Strictly speaking, this implies the total absence of any stimulus referable to male gametophyte. e.g. *Poa* spp. (Tinney, 1940; Akerberg, 1941, 1943; Engelbert, 1941; Christoff, 1942; Grun, 1955; Hakansson, 1951); *Cortaderia jubata* (Philipson, 1978).

Parthenogenic development of unreduced egg is reported in *Calamogrostis chalybaea* (Stenar, 1932; Nygren, 1946), *C. purpurea* (Nygren, 1946), *Capillipedium parviflorum* (Celarier and Harlan, 1957), *Pennisetum orientale* (Choda *et al.*, 1981), *Poa alpina* (Hakanseeon, 1943), *P. arctica* (Engelbert, 1941), *P. pratensis* (Akerberg, 1942, 1943) and *Cortaderia jubata* (Costas-Lippmann, 1979).

Autonomous development of embryo are not uncommon. Murty and Rao (1972) found the presence of a proembryo in the aposporous embryo sac even before pollination and fertilization. Autonomous development of embryo is also reported in *Poa pratensis* (Tinney, 1940), *P. alpina* (Hakansson, 1943, 1944), *Paspalum dilatatum* (Smith, 1948), *P. secans* (Snyder, 1957), *Cenchrus glaucus* (Fisher *et al.*, 1954; Shanthamma, 1982) and *Calamogrostis stricta* (Greene, 1984).

4. Eugamy

Reports of the occurrence of normal fertilization in apomictic eggs are not infrequent in literature, e.g. *Poa* spp. (Akerberg, 1939; 1942; Akerberg and Bingefors, 1953, Grun, 1954; Muntzing, 1940).

5. Androgenesis

The development of haploid embryo from the sperm nucleus is called Androgenesis. Here the egg nucleus degenerates and the sperm nucleus functions in the cytoplasm of the egg cell, thus producing a "male embryo". It is believed to occur in *Poa alpina* (♂) x *Poa pratensis* (Akerberg, 1942; Akerberg & Bingefors, 1953).

Reddy (1977) traced the evolutionary trend in apomictic mechanism. According to him, diplospory, commonly associated with the apomicts of Festucoideae, is primitive to apospory. He emphasizes the significant role of the four-nucleate unreduced embryo sac as it easily maintains chromosomal balance between the embryo, endosperm, and maternal tissue (2:3:2). In Andropogoneae, the sexual mechanism ends at the megaspore mother cell stage (MMC degenerates) and the occurrence of four-nucleate unreduced embryo sacs is predominant. In the apomicts of Panicoideae the unreduced eight-nucleate embryo sacs occur along with unreduced four-nucleate embryo sacs. This denotes the intermediate position of the Panicoideae.

Polyploidy and Apomixis

The close connection that exists between polyploidy and apomixis prompted some students of apomixis to advance the suggestion that polyploidy is a possible cause for the origin of apomixis.

In some grass genera like *Bothriochloa, Capillipedium* and *Dichanthium* (Borgaonkar, 1963) diploids are sexual, tetraploids exhibit various degrees of sexuality and higher polyploids are either completely sexual or essentially obligate apomicts. In *Dichanthium* and *Bothriochloa* (Harlan and deWet, 1975) crossing experiments have revealed that hexaploids are facultative apomicts. Absence of regular association between polyploidy and apomixis as seen in a large number of sexual polyploids and the presence of a few apomictic diploids as in *Nardus stricta 2n = 26,* (Rychlewski, 1961), *Panicum antidotale* 2n = 18 (Shamakumari, 1961), *Pennisetum ramosum* (Narayan, 1962), *P. hohenackeri* 2n = 18 (Kurup, 1973) indicate that polyploidy by itself may not be responsible for the origin of apomixis.

Norrmann *et al.* (1989) reported that *Paspalum intermedium, P. quadrifarium, P. aumani, P. brunneum, P. rufum* and *P. maculosum* have both diploid (2n = 20) and *tetraploids* (2n = 40). The diploid plants are sexual while tetraploids are facultative apomicts. In the genus *Panicum* accessions from Kenya and Tanzania studied by Combes and Pernes (1970) were shown to be either tetraploid (2n = 32) and apomictic, or diploid (2n = 16) and fully sexual. Similarly in *Themeda triandra* (Liebenberg, 1990) diploids and two of the tetraploids are sexual while penta and hexploids are obligate apomicts. In *Paspalum guaranticum* (Burson and Bennett, 1970b) diploids (2n = 40) are sexual while tetraploids (2n = 80) are obligate apomicts.

In the genus *Pennisetum* while forms with higher chromosome numbers are sexual as in *Pennisetum macrostachyum* 2n = 68 and *P. latifolium* 2n = 36 while *P. clandestinum, P. setaceum* (Narayan, 1951), *P. hohenackeri* (Kurup, 1973) and *P. mezianum* (Shanthamma, 1973) with lower chromosome numbers are apomictic. All these lead to the conclusion that in the genus *Pennisetum* polyploidy could not be a factor in the origin of apomixis. It is the genotype that determines the presence or absence of apospory and not the chromosome number as such. Stebbins (1950) is of the opinion that polyploidy may "reinforce the aciton of genes favouring apomixis".

In *Calamogrostis* (Nygren, 1948, 1951) chromosome doubling by colchicine induces change from sexuality to apomixis. A very similar situation exists in *Panicum maximum*, an important forage grass in certain tropical regions (Savidan *et al.*, 1978). Here, like in *Dichanthium,* the cytogenetics and embryology have been closely studied and extensive collections of wild material have been made. The aposporous embryo sacs are of the four-nucleated type characteristic of Panicoideae. They are easily distinguished from the sexual ones in squash preparations. Facultatively apomictic tetraploids are common. In Kenya, local diploid populations have been found, which are totally sexual. Sexual tetraploids obtained by chromosome doubling from such diploids are used in the breeding programme. Such sexual tetraploids are not able to maintain themselves in nature; they are "wiped out" by the apomicts (Pernes *et al.* 1975). Apomictic diploids obviously does not exist. Dihaploids from apomictic tetraploids are either sexual or sterile. On levels higher than tetraploidy, only apomicts are known.

Causes of Apomixis

According to Ernst (1918) hybridization is responsible for apomixis. This was criticized by Holmgren (1919) who stated that hybridity is only one of the prerequisites for apomixis. Hybridity and heterozygosity are now understood as a consequence rather than a cause of apomixis.

The genetic expression for and the ultimate development of aposporous sacs in diploids may be influenced by the environment. Seasonal environmental changes may modify the expression of apomixis and this appears to have occurred in the *Paspalum rufum*. Quarin (1986) reported a similar response between seasonal changes and the incidence of apomixis in *Paspalum cromyorrhizon*.

In facultative apomicts environmental conditions seem to play a key role in the shift from sexual mode of reproduction to apomixis. *Deschampsia caespitosa* reproduces sexually in Sweden and apomictically in California.

It has been shown in *Dichanthium aristatum* (Knox, 1967; Knox and Heslop-Harrison, 1963) that there is a tendency more apomictic embryo sacs in short days and that in long days promoted sexual reproduction. However, McWilliam *et al.* (1970) could demonstrate no relationship between photo period and apomixis in aposporous *Hyparrhenia hirta*.

Apomixis and Plant Breeding

Although apomixis has been known for many years, it is only recently that its potential to transform the agricultural scene is being widely realized. Indeed, according to one view (Jefferson, 1994), exploitation of this single trait in plant breeding can usher in the greatest change in agricultural practice since the dawn of cultivation. As to the significance of apomixis, plant breeders, employing hybridization have combined genes from diverse plants, including wild species. Neverthless it is necessary that a prized hybrid be multiplied in a way that the progeny is identical to the parent, i.e. selected hybrid. In crops, hybrids are highly prized on account of their 'hybrid' vigour, but their multiplication through seeds poses a serious problem, and is at present almost impossible, since normal sexual reproduction involves segregation and recombination of chromosomes, leading to mixed progeny. If, however, some way could be found of stopping reduction division in the megaspore mother cell and obtain embryos, say parthenogenetically without fertilization (or alternatively from nucellar cells), a particular hybrid can be 'fixed' with seeds which all form plants true to the parental type. This can be a great boon to agriculture.

Apomictic processes bypass both meiosis (apomeiosis, after Renner, 1916) and egg cell fertilization, producing offspring which are exact genetic replicas of the mother plant. Particular genotypes or complex traits as well as hybrid vigor, impossible or difficult to perpetuate through sexual reproduction, can be easily maintained and multiplied indefinitely under an apomictic mode of reproduction. Therefore apomixis would be of great value if introduced into major food crops, and would allow the propagation of heterotic hybrid combinations, leading to increased food production in developing countries (Savidan and Dujardin, 1992; Jefferson, 1994). Most major cereal crops belong to Gramineae, a family in which apomixis is widely expressed (Brown and Emery, 1958) especially in wild relatives of maize, wheat and pearl millet.

Since apomictic plants are derived from embryos that have genetic constitution of the female parent, they can greatly facilitate hybrid seed production and be valuable because heterosis will be perpetuated through successive generations via seed. Thus, a prized hybrid, once produced, can be multiplied vegetatively via apomictic seeds at no additional cost to the farmer. The hybrid will not lose its genetic make up—unavoidable in sexual crossing—and will breed true. The farmer can also be spared incurring expenditure on purchasing expensive hybrid seeds which normally he must procure every year from seed companies (specializing in maintaining male sterile and restorer lines and related techniques) and which is on average atleast 10 times costlier. It is this attractive prospect that has awakened world-wide interest in introducing of Apomixis in crops (Bashaw, 1980; Hanna and Bashaw, 1987; den Nijs and Van Dijk, 1993; Khush, 1994; Hanna, 1995; Vielle *et al.,* 1996).

Crossing experiments between sexual crops and related wild species have been performed in recent years. Such experimentas have given valuable theoritical information and proved the possibility to transfer resistance from the wild species. Moreover they have shown that the introduction and manipulation of apomictic reproduction in sexuals is no impossibility.

Wheat

Wheat (*Triticum aestivum*) can be crossed with the apomict *Agropyron scabrum.* It is uncertain, if this possibility can be practically exploited. It may be mentioned that pseudogamy has been reported in intergeneric hybrids between *Triticum* and *Avena* (Kruse, 1969). According to the author, the mode of reproduction in this case depended either on the intrageneric hybridization or on the UV-irradiation preceding pollination.

Induced apomixis in wheat, *Triticale* and wheat-rye hybrids has been reported by Russian scientists. Chistyakova (1978) obtained induced apomixis in wheat after pollination with Colchicine-treated pollen. Kandelaki (1978) obtained both haploid and diploid parthenogenesis in F_1 hybrids between wheat and rye after species-alien pollination. Selection of plants giving a high rate of maternal diploids and/or haploids in distant crosses would be one step towards producing a synthtic apomict.

Mujeeb Kazi (1996) reported apomixis in trigeneric hybrids. In this study a self-sterile F_1 hybrid of *Triticum aestivum / Leymus racemosus* (2n = 5x = 35) upon pollination by *Thinopyrum elongatum* (2n = 10x = 70) yielded derivatives that possessed 35 to 70 chromosomes. The single 35 chromosome derivative is considered to be the product of parthenogenetic egg-development (apomictic) while the 70 chromosome derivatives resulted from the fertilization of a 35 chromosome egg cell with pollen from *Th. elongatum* with 56 chromosome back cross I *T. aestivum / L. racemosus / T. aestivum* plants were pollinated by *Th. elongatum* trigeneric derivatives of normal chromosomal composition.

Maize

In Maize, several genes are known to affect the reproductive behaviour, for instance by influencing meiosis. It would not be unbelievable to produce a functioning apomictic system by the help of induced mutations. So far, more interest has been attached to crosses with apomictic forms of *Tripsacum dactyloides*.

The chromosome number of diploid maize is 2n = 20 and of apomictic *Tripsacum dactyloids* 2n = 72 (sexual forms of the later species have 2n = 36). By crossing diploid or tetraploid maize with *Tripsacum* and by back-crossing, hybrids have been obtained which represent a variety of combinations between the genomes of the parental species.

Plants with 36 *Tripsacum* and 20 *Zea* chromosomes give a very variable off-spring after pollination with maize. Among others, partially apomictic plants with the chromosome numbers 2n = 28 (dihaploid), 38 and 56 occurred (Engle *et al.* 1974).

The occurrence of "Pseudo apomixis" in certain hybrdis with 2n = 46 should be mentioned (Harlan *et al.* 1978). In the meiosis, often 18 bivalents and 10 univalents are formed. The latter (= the maize chromosomes) are eliminated in the first meiotic division. After the second division, no cell wall is formed, and the egg cell retains the 36 *Tripsacum* chromosomes. A back cross with tetraploid maize restores the chromosome number 2n = 46. This process has been repeated for several generations.

Resistance genes have already been transferred from *Tripsacum* to *Zea*. To obtain high-yielding apomicts by *Zea-Tripsacum* hybridization remains a difficult task. Petrov *et al.* (1973) claim to have transferred "genes, controlling apomictic elements, from the chromosomes of *Tripsacum* to those of maize". This could be an important step towards the introduction of apomixis into cultivated maize.

The ORSTOM-CIMMYT project aims at transferring apomixis from *Tripsacum dactyloides* to maize. F_1 progenies from the wide crosses of these two species have been back crossed twice to maize and apomietic back cross progenies with 20 chromosomes of maize and 2–6 chromosomes of *Tripsacum* have been selected. Further backcrosses are underway (Savidan *et al.*, 1994).

Kindiger *et al.* (1996) reported that an apomictic form of hybrid maize would provide an immortalized line which would be stabilized against genetic change. The development of agronomically superior, apomictic maize hybrids could provide a superior level of food security in developing nations as well as altering commercial and public maize breeding programmes - systems in developed nations. Back cross selections obtained from an apomictic 38-chromosome (20 maize + 18 *Tripsacum*) maize-*Tripsacum* hybrid have resulted in the development of apomictic, 39 chromosome individuals with 30 maize + 9 *Tripsacum* chromosomes. The identification of these materials advances two major objectives (i) the elimination of nine *Tripsacum* chromosomes which do not possess gene(s) controlling apomictic reproduction and (ii) the continued refinement of an apomictic maize line. This study was conducted to determine whether the 30 chromosome materials reproduce by apomixis. The generation of these materials could result in the eventual development of a hybrid maize with an apomictic reproductive system (Kindiger *et al.*, 1996).

Forage Grasses

An interesting project aiming at the development of new apomictic forage grasses is carried out at Gatersleben by Grober *et al.* (1974, 1976). According to the authors, spontaneous hybridization between species or genera is a pre-requisite for the origin of apomixis within a plant group, vigorous, but more or less sterile, hybrids may be propagated vegetatively, until by several steps of evolution and natural (or human) selection a functioning apomict may arise.

Grober *et al.* (1974) performed generic and specific crosses between species belonging to the genera *Bromus, Festuca* and *Lolium*. More than 100 crossing combinations were realized and maternal plants of different origin were used to enhance the formation of hybrid off-spring. In many cases, embryo culture was necessary to get adult hybrids. The hybrids obtained were in many cases sterile but often very vigorous.

To date only an apomictic forage grass species has been improved through breeding. Bashaw (1980) developed three improved cultivars of apomictic buffel-grass through hybridization between sexual and apomictic clones.

Thus apomixis could be utilized to produce high yielding varieties with desirable genetic traits with much ease than through the routine breeding methods. Gupta (1968) said that in most of the agamic complexes in grasses and particularly in Andropogoneae, diploids are sexual, tetraploids are facultative apomicts while pentaploids and hexaploids are obligate apomicts with some exceptions.

Grimanelli *et al.* (1998b) reported that in the *Tripsacum* agamic complex, all polyploids reproduce through diplosporous type of apomixis and diploids are sexual. They used molecular markers linked with diplospory to analyse various generations of maize—*Tripsacum* hybrids and back cross derivative to derive a model for the inheritance of diplosporous reproduction. The results suggest that the gene or genes controlling apomixis in *Tripsacum* are linked with a seggregation distorter type system promoting elimination of the apomixis allelss when transmitted through haploid gametes. Hence this model offers an explanation of the relationship between apomixis and polyploidy (Grimanelli *et al.* 1998b).

Pearl Millet

Morgan *et al.* (1998) transferred the aposporus apomictic mode of reproduction from *Pennisetum squammulatum* into pearl millet (*Pennisetum glaucum*) through a trispecific double-cross hybrid that was subsequently back crossed (BC) to tetraploid (4x) pearl millet. A partially male-fertile obligate apomictic plant, was selected in the BC_3 generation. A negative feature of this apomietic hybrid is the low seed set on open- or cross pollinated inflorescences. Historical analysis of ovaries revealed that endosperm cells degenerated 4 days after pollination in upto 80% of the ovules.

Genetic Basis of Apomixis

Apomictic species are poor subjects for genetic study. The data from most crosses between apomictic and sexual individuals have not been conclusive. Some of the difficulties are due to the complex polyploid nature of apomictic species. Taliaferro and Bashaw (1966) reported that aposporous apomixis in buffelgrass is controlled by two genes with epistasis. Somewhat similar results were obtained from studies with bahiagrass by Burton and Forbes (1960) who postulated recessive genes for apomixis. Funk and Han (1967) postulated apomixis to be controlled by two or more dominant genes in *Poa pratensis*. Hanna *et al.* (1973) found sexuality to be dominant in crosses of sexual X apomictic *Panicum maximum*. Savidan (1980, 81) reported that apospory in guinea grass (*Panicum maximum*) is controlled by a dominant gene. Dujardin and Hanna (1983) suggested apospory in *Pennisetum squammulatum* is under dominant gene control. Thus, the available data suggests that only a few genes control apomixis with profound effects on the mechanism of sexual reproduction.

S.C. Maheshwari *et al.* (1998) has discussed in detail genetic basis of apomixis. Genetic basis of apomixis is of great significance for any research aiming to transfer 'apomixis' to nonapomictic plants (Hanna, 1995), since if many genes are involved, the apomixis trait will be difficult to engineer. Broadly, three sets of genes appear to be involved in female meiosis and reproduction for: (i) the progress of normall meiosis, (ii) the mitosis of megaspore nucleus followed by embryo sac development and (iii) developmentof the embryo. Typically, it is the failure of meiosis during megasporogenesis and the initiation of mitosis in the unfertilized diploid egg which result in apomixis. However, in apospory and adventive embryony, renewed mitotic activity of nuclei in nucellar cells is responsible for apomixis. The precise genetic basis of apomixis may be different in various types of apomixis.

A consideration of various biochemical events underlying mitosis and meiosis indicates that scores and even hundreds of genes must be involved in these processes and, on account of the existence of various types of apomixis, the deciphering of the mechanisms of apomixis may seem a herculean task, let alone engineering such a trait. But, fortunately, the basic genetics is not as complex because, generally, there are critical nodal points in development and key regulatory genes for these modal points are few. Given the knowledge of existence of homeotic genes in animals and lately also in plants, one may speculate that for initiating each of the process (i), (ii) and (iii) mentioned above, there may be a separate regulatory gene of this kind. Although the genetics of apomixis has not been thoroughly investigated, it does appear that only one or two genes may control this trait (Nogler, 1984; Den Nijs and Van Dijk, 1993). Atleast in *Pennisetum* and *Panicum*, apomixis is controlled by a single gene or locus. Neverthless, independent evolution of apomixis has, apparently, led to some diversity in the mechanism of apomixis even in plants that show the same type of apomixis (e.g. diplospory). Thus although in certain plants apomixis appears to be governed by only one gene and it is inherlted as a dominant trait, in some other plants it has been reported to be recessive, pointing to the involvement of atleast two types of genes (S.C. Maheshwari *et al.*, 1998).

Construction of linkage maps for the chromosome controlling diplosporous apomixis in *Tripsacum*, a wild relative of maize, was carried out in both tetraploid-apomictic and diploid-sexual species using maize restriction fragment length polymorphism (RFLP) probes by Grimanelli *et al.* (1998). A high level of colleaniarity was observed between the *Tripsacum* chromosome carrying the control of apomixis and a duplicated segment in the maize genome. In the apomictic tetraploid, there was a strong restriction to recombination, as compared to the corresponding genomic segment in sexual plants and maize. This suggests that apomixis, although inherited as a single mendelian allele, might really be controlled by a cluster of linked loci. The analysis by Grimanelli *et al.* (1998) also revealed the tetrasomic nature of the inheritance of the chromosomal segment controlling apomixis, which contradicts the usually accepted hypothesis of an allopolyploid origin of apomictic species.

Apomixis has a genetic basis, but it is still a matter of question how it is regulated. However, important progress has been made in taxa like *Panicum maximum* (Savidan, 1982a): where results are in agreement with a monogenic system of control. What remains obscure is how a single gene regulates, in aposporous apomicts, megaspore abortion, the formation of embryo sacs from nucellar cells, and the parthenogenic development of unreduced egg cells from aposporous embryo sacs. Is parthenogenesis a consequence of apospory or are both of them controlled by two linked genes? Savidan (1982a) hypothesized that parthenogenesis in aposporous embryo sacs is consequence of the timing in the embryo sac development. Meiotic embryo sacs reach maturity by the time of anthesis, where as aposporous sacs seem to have a precocious development. Thus, pollen tube penetration into the pistil is synchronized with meiotic embryo sac maturity. However, the pollen tube penetration into ovule-bearing aposporous embryo sacs may be too late for fertilization, because the sacs have already reached maturity. In this way, parthenogenesis may take place as a consequence of failure of fertilization. Following this hypothesis, it may be possible that forced precocious pollination results in fertilization of the unreduced egg cell of an aposporous embryo sac.

Introgression of Apomixis and Mapping the Apomictic Genes

S.C. Maheshwari *et al.* (1998) in an excellent review has given various aspects of mapping the apomictic gene. Because only a few genes (probably a single key gene) seem to be involved in the control of apomixis, many breeders have been encouraged to introgress the apomictic trait in crops from wild relatives. Although almost all the studies todate have relied on classical and empirical breeding methods, such as repeated back-crossing by a donor or the gene, some notable progress has been made using these methods. Example wise, the aposporous trait found naturally in *Pennisetum squammulatum* is being introduced into *Pennisetum glaucum*, a cultivated species by Ozias-Akins and co-workers (1993). One resulting line has the apomictic gene located on a supernumerary *squammulatum* chromosome. Fortunately, RFLP markers are now available for several important crop plants and efforts are being made also to develop RFLP maps. Two molecular markers co-segregate with apomixis, thereby giving a valuable opportunity to locate the apomictic gene and work on this is being actively pursued by this group in USA. Similar efforts are underway to transfer the apomictic trait also across genera, for example in Maize (Leblanc *et al.*, 1995) from *Tripsacum*, a wild relative, and in wheat (Liu *et al.*, 1995) from *Elymus* as well as to isolate gene for apomixis. According to S.C. Maheshwari *et al.* (1998) there is some progress towards transferring apomixis from wild to cultivated species and also identifying gene loci responsible for apomixis. Nonetheless, so far, the identity of the genes or of the products and their function(s) are unknown and this is clearly a task for the future.

The question remains whether the two essential components of apomictic reproduction, followed by parthenogenesis, are determined by independent or linked genetic factors (Leblanc *et al.* 1995). Indeed, there is evidence that apomeiotic hybrids lacking parthenogenetic capacities are rarely encountered (Nogler, 1984; Kojima *et al.,* 1994). Moreover, in *Tripscum* species, as in most apomicts, the presence of gene(s) responsible for apomeiosis generally results in maternal progenies (Nogler, 1984; Asker and Jerling, 1992). This may be interpreted in two ways: either the region controlling diplospory contains a major factor with pleiotropic effects on the other essential components of gametophytic apomixis, or it contains several tightly linked factors each controlling one simple component. The risk of selecting purely apomeiotic or diplosporous genotype in derivatives of *T. dactyloides* by using markers co-segregating with diplospory would therefore, be proportional to the strength of this linkage (Leblanc *et al.*, 1995).

Leblanc *et al.* (1995) studied the apomictic behaviour in natural and artificial tetraploids of *Tripsacum*. A collection of embryogenic diploid calli of *Tripsacum* was established and treated with colchicine to induce chromosome doubling. Sections containing duplicated cells in calli were identified using flow cytometry and ploidy level was determined in the regnerated plantlets. Tetraploid plants from several origins were obtained. In contrast to wild polyploid plants, which show apomictic development, the regenerated tetraploid plants reproduced sexually. By hybridizing these plants with wild tetraploid apomicts, various populations were established; these will allow a study of the inheritance of apomixis in *Tripsacum* (Leblanc *et al.*, 1995).

A cultivated member of the genus, pearl millet (*Pennisetum glaucum*), reproduces sexually. A wild relative of pearl millet, *Pennisetum squammulatum*, that is an obligate aposporous species, is cross-compatible with pearl millet when use as a pollen donor in the inter-specific cross (Ozias-Akins *et al.*, 1998). Ozias-Akins *et al.* (1998) presented the genetic mapping of 13 molecular markers in an interspecific hybrid population of 397 individuals that segregates for apomixis and sexuality. They found that 12 out of 13 markers strictly cosegregated with aposporous embryo sac developed, clearly defining a contiguous-apospory-specific genomic region in which no genetic recombination was detected. Lack of or suppression of recombination may be coincidentally associated with the chromosomal context of the apomixis locus or it may be a consequence of its evolution that is essential for preservation of gene function as has been shown in studies of complex loci in both plant and animal species.

ANNEXURE

ADDRESSES OF SCIENTISTS WORKING ON APOMIXIS IN GRASSES

1. **Dr. E.C. Bashaw**
Department of Rangeland Ecology and Management
Texax A&M University
College Station, Texas 77843-2474
USA

2. **Dr. Ravindra K. Bhanwra**
Department of Botany
Panjab University
Chandigarh 160 014
INDIA

3. **Dr. Byron L. Burson and Dr. Paul W. Voigt, USDA-ARS**
Department of Soil & Crop Science
Texas A&M University
College Station TX 77843-2474 - USA
E-Mail: n-burson @tamu.edu

4. **Dr. J.G. Carman**
Department of Plants, Soils and Biometerology
Utah State University
Logan UT 84322-4820
USA
E-mail: jcarm@mendel.usu.edu

5. **Dr. Charles Crane**
Department of Soil and Crop Sciences
Texas A&M University
College Station TX 77843-2474
USA
E-mail: monosom @tamu.edu

6. **Dr. Daniel Grimanelli**
ORSTOM-CIMMYT
Apdo 6-6 41
Mexico DF 06600
Mexico
E-mail: dgrimanelli@cimmyt.mx

7. **Dr. Diego Gonzalez De Leon**
Applied Molecular Biology
CIMMYT
Apdo-Postal 6-641
Mexico DF 06600
Mexico

8. **Dr. Cacilda Do Valle**
EMBRAPA-CNPGC
Caixa postal 154
Campo Grande, 79002-970 MS
Brazil

9. **Dr. G. Evers**
Texas A&M University Agricult. Research Station
Angleton Texas 77515
USA

10. **Dr. Wayne Hanna**
USDA - ARS-SAA
Coastal Plain Experimental Station
P.O. Box 748
Tifton GA 31793
USA
E-mail: forage@tifton.cpes.peachnet.edu

11 **Dr. Mark Hussey**
Department of Soil and Crop Sciences
Texas A&M Univeristy
College Station TX 77843-2747
USA

12. **Dr. Elizabeth A. Kellog**
Harvard University Herbaria
Divinity Avenue
Cambridge MA 02138
USA

13. **Dr. Oliver Leblanc**
OSTOM-LRGAPT
Labororatoire de Resources Genetique et d' Ameliorationdes plantes
UMR 9938 CNRS- INRA-ENS
46 Allee de Italie
LYON CEDEX 07
FRANCE
E-mail: leblanco@orstom.fr

14. **Dr. E.L. Lubbers and Dr. L. Arthur**
Department of Horticulture
University of Georgia
Coastal Plain Experimental Station
Tifton, GA 31793
USA

15. **Dr. P. Ozias - Akins**
USDA-ARS
University of Georgia
Coastal Plain Experimental Station
P.O. Box 748
Tifton GA 31793-0748
USA

16. **Dr. M.D. Peel**
Department of Plant Sciences
North Dakota State University
Fargo ND 58105
USA

**17. Dr. Camilo L. Quarin and
Dr. G.A. Norrmann**
Instituto de Botanica del Nordeste
Facultad de Cincias Agrarias
Universidad Nacional del Nordeste
Castilla de Correo 209
3400 Corrientes
ARGENTINA
E-mail: camilo@compunort.com.ar

18. Dr. Y. Savidan
ORSTOM-CIMMYT
Apdo-6-641
06600, Mexico DF

**19. Prof. A.N.R. Sindhe, Prof. M.
Muniyamma and Prof. C. Shanthamma**
Department of Botany
University of Mysore
Manasagangothri
Mysore - 570 006
INDIA

20. Dr. A.K. Srivastava
Department of Botany
C.C.S. University
Meerut 25004
INDIA

Identifying Apomixis

Apomixis of various types and at various frequencies would probably be found in more plant species if research efforts were concentrated in this area (Hanna and Bashaw, 1987). Unfortunately, it is easy for plant breeders to overlook possible sources of apomixis due to a lack of understanding of the apomictic mechanisms. Failure to obtain F_1 plants from a cross or production of uniform progeny in an F_2 population may be blamed on poor emasculating or crossing technique. Although absence of F_1 plants in crosses and plant uniformity in F_2 population do not guarantee apomictic reproduction, they are indicators of apomixis that should be evaluated.

There is no single method or technique for identifying and confirming apomixis. A combination of cytological, genetic and progeny tests can be used to detect apomixis, identify the apomictic mechanism, and determine the frequency of apomictic reproduction. The following discussion will concentrate on methods for detecting apomixis in sexual species because the greatest impact of apomictic reproduction would be to use it in the breeding of economically important cereals, fruits and vegetables. The same general principles would apply for detecting sexuality in obligate apomictic species (Bashaw, 1962; Hanna *et al.*, 1973).

Breeding Behaviour

(A) Uniform progeny from a heterozygous plant or an open-pollinated species is one of the best indicators of obligate apomixis. Obligate apomixis will result in uniform progeny identical to the maternal parent on which the seed was produced. Progeny testing of open-pollinated seed from cross-pollinated species is a quick and easy way to screen a large number of introductions.

(B) Production of maternal phenotypes in crosses is another indicator of apomixis but further studies are needed to eliminate the possibility that the crossing technique might have allowed selfing. This would be especially important to check in self-pollinated species or in homozygous lines.

(C) Use of a dominant marker gene in a male pollinator carefully crossed onto various accessions of plants suspected of reproducing by apomixis but without the marker gene allows both detection and confirmation of apomixis. Production of maternal progeny with the recessive gene indicates apomixis. This technique is more efficient if a marker gene identifiable as early as the seedling stage is used.

(D) Unusually high seed set and/or uniform maternal progeny in plants with irregular chromosome numbers or behaviour such as in wide crosses, aneuploids and odd polyploids may indicate apomictic reproduction. In genotypes that are similar except for reproduction behaviour, apomicts will usually set more seed than sexual plants, provide viable pollen is available.

Apomictic reproduction eliminates the meiotic process which can produce unbalanced and lethal female gametes.

Seed and Seedling Characteristics of Apomicts

Multiple seedlings per seed, seed with two or more fused ovaries, multiple stigmas and multiple ovules per floret (Hanna *et al.* 1970, 1973) may indicate apomixis. Ovaries/pistils with multiple stigmas and multiple ovules per floret often result in seed with fused ovaries. Multiple seedlings per seed can be due to: (1) the development of multiple aposporous embryos develop in both sexual and apomictic embryo sacs in the same ovule (one or more seedlings may be morphologically different than the rest), or (3) development of embryos in two separate sexual embryo sacs in an ovule (plants may be morphologically different). Cytological studies are needed to confirm these characteristics.

Cytological Tests

Cytological studies are useful for detecting apomixis and are essential for determining the apomictic mechanism.

Apospory

Multiple embryo sacs clustered together and/or randomly distributed throughout the ovule, usually four or fewer nuclei per embryo sac, and lack of antipodals in the embryo sac at anthesis indicate apospory. Single aposporous embryo sacs also are possible and can be detected by their lack of antipodals. The lack of antipodals in a mature embryo sac is one of the most reliable distinguishing characteristic of apospory.

Diplospory

Cytological studies to detect diplospory must be conducted between the stage when the megaspore mother cell differentiates and early development of the embryo sac. This stage is synchronized with microsporogenesis in many species. The lack of meiosis and chromosome reduction (may be difficult to determine) and the lack of a linear tetrad of spores resulting from meiosis (more easily determined) are characteristic of diplospory. In one type of diplospory the megaspore mother cell goes through one mitotic division to produce a dyad of microspores of which one cell functions and the other degenerates (Gustafsson, 1946; Nogler, 1984). This degenerating disomic megaspore should not be confused with the three degenerating micropylar spores in sexual reproduction. A mature diplosporous embryo sac may look identical to a mature sexual embryo sac at anthesis.

Adventitious Embryony

Adventitious embryony can be identified by the pressure of proembryos that develop directly into embryos from somatic cells in the ovule (Bashaw, 1980a) No apomictic embryo sacs are formed in adventitious embryony; however, sexual embryo sacs may be formed in the same ovule. The sexual embryo sacs are needed for the formation of endosperm.

Cytological Technique

Systematic examination of anatomical features throughout development of an ovule is needed to thoroughly understand the cytological basis for the various reproductive mechanisms (Bashaw, 1980a). Embedding ovules in paraffin or a resin, sectioning and staining with safranin-fast green as outlined in Gurr (1965) as well as other plant microtechnique books is satisfactory for studying general morphological detail. Chromtin stains are needed for observing chromosome detail. In histological studies, care should be taken to properly align the ovules so that longitudinal sections through the gametophytic tissue are obtained (Bashaw, 1980a).

Clearing methods using aromatic esters that greatly reduce the time needed to prepare samples before observations have been reported recently (Young *et al.*, 1979; Crane and Carman, 1987). These technique allow one to use interference contrast microscopy to observe the contents of the entire ovule by changing the focal level. These methods make possible the determination of the reproductive mechanism in 2 or 3 days after ovule collection.

CHAPTER 9

POLYEMBRYONY

Polyembryony has been defined as the occurrence of more than one embryo in a seed (Maheshwari, 1950). Occurrence of polyembryonic seeds in sexullay reproducing species of Gramineae is a very rare phenomenon. However the apomictic grasses produce twin and triple embryos more frequently. Schrenk (1894) for the first time illustrated and briefly described the presence of two plumules and two primary shoots and single cotyledon in two kernels of maize. Kemption (1913) described similar seeds in maize. Zimmerman (1904) reported that seeds of *Poa pratensis* sometimes contain from one to three embryos. Zinn (1904) also noted polyembryonic seeds in *Poa pratensis, P. nemoralis* and *P. compressa.* Weatherwax (1921) and Kiesselbach (1925) found and described several cases of false polyembryony in maize. Since then, polyembryony has been reported in several members of Poaceae by several embryologists like Nishimura (1922), Anderson (1927), Armstrong (1937), Muntzing (1937, 1940), Akerberg (1939), Tinney (1940), Engelbert (1941), Smith (1948), Warmke (1954), Fisher *et al.,* (1954), Snyder *et al.* (1955), Farquharson (1955), Venkateswarlu and Devi (1964), Reddy and D'Cruz (1969), Gupta and Yashvir (1971), Shanthamma and Narayan (1977), Das and Islam (1977), Choda and Bhanwra (1977, 1980), Sindhe *et al.* (1980), Bhanwra *et al.* (1982), Shanthamma (1982), Bhanwra (1988), and Febulaus and Pullaiah (1991). (Table III gives details of polyembryony).

Polyembryony can be broadly classified into "simple" and "multiple" depending upon the presence of one or more embryo sacs in the same ovule (Lakshmanan and Ambegaokar, 1984). Simple polyembryony can be sexual or asexual. The examples of sexual polyembryony are the embryos originating from the egg cell and synergid, supernumerary embryos may also arise from proembryo. Asexual embryos develop within the embryo sac without fertilization. If the embryos develop from nucellar cells without the interpolation of the gametophytic phase, the polyembryony is said to be "adventive" or "sporophytic". In multiple polyembryony accessory embryos are produced from two or more embryo sacs in the same ovule.

SIMPLE POLYEMBRYONY

i) Cleavage Polyembryony

Development of additional embryo due to the cleavage of zygote or proembryo in less frequent in Poaceae. It has been recorded only in very few members like *Poa pratensis* (Nishimura, 1922a), *Zea mays* (Randolph, 1936), *Dicanthium annulatum* (Reddy and D'Cruz, 1969b) and rarely in *Cenchrus ciliaris* (Gupta and Yashvir, 1971).

ii) Synergid Polyembryony

Synergid origin of extra embryo within the single embryo sac has been reported in several members of Poaceae. This is evident from the occurrence of twin embryos which are surrounded by copiously formed endosperm (Sindhe *et al.*, 1980). Embryos may arise either from the fertilized synergid, or from the unfertilized synergid. In *Pennisetum dubium* the synergid autonomously develops into an embryo (Gildenhuys and Brix, 1959). Synergid polyembryony has been recorded in *Tripsacum dactyloides* (Webber, 1940), *Saccharum spontaneum, Triticum speltatum,* hybrids of *Avena strigosa* × *A. fatua,* and *Calamogrostis obtusa* × *C. purpurea* (Webber, 1940; Johansen, 1950).

Shobha and Sindhe (1987) observed sporadic occurrence of twin embryos in *Pennisetum pedicellatum*. The location and shape of the extra embryo in the embryo sac reveal that it is of synergid origin. The synergid embryo is comparatively smaller than the sexual embryo and is pear-shaped, having the contour of synergid itself. In contrast the zygotic embryo has a typical multicellular squat suspensor characteristic of all sexual embryos in Gramineae. In position too the extra embryo lies in the same locus as that of the synergid.

In *Pennisetum squammulatum* (Sindhe *et al.*, 1980) the diploid synergid embryo develops along with the zygotic embryo. The suspensor of the zygotic embryo is typically multicellular as in the Gramineae, and that of the synergid is partly uniseriate and partly biseriate.

iii) Antipodal Polyembryony

Antipodals are usually ephemeral in nature and they get degenerated. However in *Paspalum scrobiculatum* (Naryanaswami, 1954) the antipodals rarely develp into embryos. However his observations need conformity.

iv) Embryos from Endosperm

Muniyamma (1977) reports that *Brachiaria setigera* (2n = 36) is a facultative apomict. In this species polyembryony is due to formation of embryos from egg.

synergids and endosperm. Six aposporous embryo sacs from 675 post-fertilized ovules contained embryos which had originated from endosperm. Meristematic activity in localized areas of the mature endosperm produce globular or spindle-shaped structures. Enlargement and organisation of these structures led to viable embryos. The sequence of events beginning with a single cell and followed by a complete organized embryo was not observed. Cytological studies of root tips of the twin and triplet seedlings showed combination of diploid-diploid, diploid-triploid and diploid-diploid-triploid. Only two triploids were obtained from 200 seeds germinated.

According to Johri *et al.* (1992) the above report needs confirmation. The globular and spindle-shaped structures arising from mature endosperm appear (to them) like tumors.

v) Adventive or Nucellar Polyembryony

Embryos arising from the cells lying outside the embryo sac i.e. either from nucellus or from integuments are defined as adventive embryos or nucellar embryos. Nucellar polyembryony is known to occur in some members of the family. They arise due to the proliferation of the nucellar cells near the micropylar region. These can be distinguished from the adjacent nucellar cells by their larger size, dense cytoplasm and prominent nuclei.

In *Eragrostiella bifaria* Venkateswarlu and Devi (1964) recorded as many as 4 well developed embryos with suspersors in 50% of the ovules. Sonpipare (1984) reported that these embryos arise from the nucellar epidermis. Febulaus and Pullaiah (1991) also observed nucellar embryos in the same species (Fig. 23A–E) which are in conformity with the earlier reports. Nucellar embryos are also recorded in *Iseilama anthephoroides, Perotis hordeiformis, Eriochloa procera* (Venkateswarlu and Devi, 1964), *Pennisetum mezianum* (Shanthamma and Narayan, 1977) and *Cenchrus ciliaris* (Fig. 23F) (Shanthamma and Narayan, 1977; Febulaus and Pullaiah, 1991).

These nucellar embryos normally arise from the cells of the nucellus and enter the embryo sac cavity and there they complete their further development as in *Eragrostiella* and *Perotis* or they do not enter the embryo sac as in *Pennisetum* and *Cenchrus* (Shanthamma and Narayan, 1977). Das and Islam (1977) also reported that the nucellar embryos in *Cenchrus pennisetiformis* and *C. setigerus* and these have arisen from the nucellar epidermis.

These adventive embryos can be distinguished from the normal zygotic embryos by their lateral position, irregular shape and lack of suspensor. However Venkateswarlu and Devi (1964) reported nucellar embryos with suspensor in *Eragrostiella bifaria.*

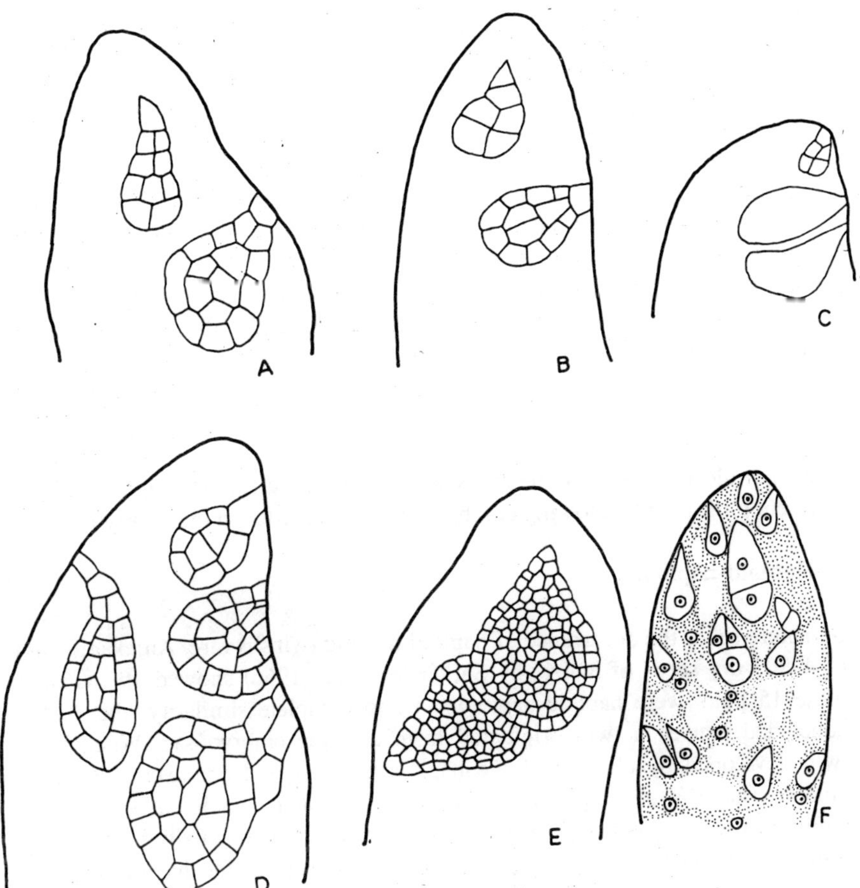

Fig. 23. Development of nucellar embryos. **A-E**. *Eragrostiella bifaria* (Febulaus & Pullaiah, 1992 b). **F**. *Cenchrus ciliaris* (Febulaus & Pullaiah, 1995).

MULTIPLE POLYEMBRYONY

Development of plural embryos from different embryo sacs of the same ovule has been reported in several members. Kiesselbach (1926a,b) reported some cases of false polyembryony in maize. Anderson (1927) reported that in *Poa pratensis* and *P. compressa* often twin embryo sacs are developed, which results in twin embryos. Development of plural embryos due to the functioning of more than one embryo sac in the same ovule has been recorded in *Oryza sativa* (Ramaiah *et al*., 1935), *Poa pratensis* (Armstrong, 1937; Tinney, 1940; Nielsen, 1946), *Pennisetum ciliare* (Fisher *et al*., 1954), *Paspalum dilatatum* (Smith, 1948), *Dicanthium annulatum* (Reddy and D'Cruz, 1969b), *Cenchrus ciliaris* (Gupta and Yashvir, 1971), *Spinifex littoreus* (Lakshmanan and Jayalakshmi, 1980), *Capillipedium huegelli* (Choda and Bhanwra, 1980) etc.

In multiple embryo sacs within an ovule, whereas the egg cell (diploid) of the unreduced embryo sac may develop parthenogenetically, the egg cell (haploid) of the reduced embryo sac may develop after fertilization forming diploid-diploid twin. If the unreduced egg is fertilized and the reduced egg develops parthenogenetically, the twin would be triploid-haploid.

Asker (1979) refers to four possible modes of reproduction in facultative apomicts: (1) the unreduced egg cells, by parthenogenesis, produce uniform maternal offsprings, (2) unreduced egg cells, by fertilization give rise to autotriploids (U hybrids), (3) the reduced egg cells by parthenogenesis, give rise to (poly) haploids, and (4) the reduced egg cells, through fertilization produces variable sexual offspring (R hybrids).

TWINS

The multiple embryos within the seed may be of different origin and with different ploidy. The following combinations are recorded in the family Poaceae.

1. Haploid-haploid

In mature seeds the occurrence of haploid-haploid twins is rare. An examination of 49903 dry seeds of maize, by Sarkar and Coe (1966) showed 49 twins. Of these 15 pairs were haploid-haploid. Due to complete similarity, the authors concluded that each twin originated from a single embryo sac. Whether the twin developed by clevage of the haploid embryo or from two cells of the embryo sac was not determined. However, as each constituent of the twin embryo in seeds, and many twin seedlings, were developmentally identical and mutually oriented, the authors presumed that clevage of the haploid embryos resulted in twins.

2. Haploid-diploid

In Haploid-diploid twins the haploid embryo develops from the unfertilized synergid and the diploid from the fertilized egg. In *Triticum vulgare* Yamamoto (1936) reported that the haploid embryo arises from the unfertilized egg and the diploid embryo from the fertilized synergid. Haploid-diploid twins were also reported in *Triticum durum* (Kihara, 1936), *T. vulgare* (Kasparayan, 1938, Krishnaswamy, 1939), *Secale cereale* (Kostoff, 1939), *Dactylis* (Nissen, 1937), *Poa* (Skovsted, 1939) and *Phleum* (Muntzing 1937).

3. Haploid-triploid

Haploid-triploid twins were reported only in *Phleum* (Muntzing, 1937).

4. Diploid-diploid

Diploid-diploid twins can arise by fertilization of a synergid and egg cell as in *Pennisetum squamulatum* (Sindhe *et al.*, 1980). In *Poa pratensis* (Nielson, 1946) diploid-diploid combinations are considered to arise from two somatic cells or from one egg cell and synergid of a single embryo sac that develops parthenogenetically. These twins can also arise from the cleavage of the proembryo as in *Zea mays* (Randolph, 1936), *Poa pratensis* (Nishimura, 1922a,b), *Dicanthium annulatum* (Reddy and D'Cruz, 1969b) and *Cenchrus ciliaris* (Gupta and Yashvir, 1971).

5. Diploid-triploid

In diploid-triploid combination, the egg of one embryo sac develops parthenogenetically whereas that of the second develops as a result of fertilization by a male gamete carrying the reduced chromosome number as in *Poa pratensis* (Nielsen, 1946). Yamamoto (1936) recorded diploid-triploid twin in *Triticum vulgare*. Such twins are also reported in *Secale cereale* (Kostoff, 1939). According to Yamamoto (1936) the triploid embryo might have originated by the fertilization of an unreduced (aposporic) embryo sac or from the fusion or a cell of haploid embryo sac with two male gametes or one unreduced male gamete, while the diploid embryo develops from the zygote. They also stated that the triploid embryo might have arisen from a part of endosperm but they didn't show any positive evidence as there is no authentic report of endospermic origin of embryos. Diploid-tripoid twins have also been reported in *Avena* (Muntzing, 1937), *Phleum* (Muntzing, 1937, Skovsted, 1939), *Lolium* (Muntzing, 1937), *Dactylis* (Skovsted, 1939), *Secale* (Muntzing, 1937) and *Poa* (Muntzing, 1937, Nissen, 1937; Skovsted, 1939).

6. Diploid-tetraploid

In *Poa pratensis* (Nielson, 1946) the diploid member originates from the zygote. The tetraploid member is formed by the parthenogenetic development of the egg with a somatic doubling of the chromosomes of the nuclear compliment at the time of the zygote division, or fertilization with an unreduced or restitution male nucleus or with a high polyploid gamete. However, he considered origin of tetraploid embryos involving gametes to be uncommon and difficult to comprehend. Diploid-tetraploid twins have also been reported in *Triticum* (Namirawa and Kawakami, 1934; Yamazaki, 1937).

7. Triploid-triploid

Nielson (1946) recorded the formation of triploid-triploid twin in *Poa pratensis.* According to him these combination may form by the fertilization of the two cells. Triploid-triploid twins have also been reported in *Dactylis* (Skovsted, 1939) and *Lolium* (Muntzing, 1938).

TRIPLETS

Diploid-diploid-triploid

Triplets are of rare occurrence in Poaceae. Muniyamma (1977) reported triplets in *Brachiaria setigera.* Out of 200 seedlings germinated two are triploids. These were found to be diploid-diploid-triploid.

Apomixis and Polembryomy

In *Poa pratensis and Elymus rectisetus* characterised by the gametophytic apomixis, the embryo of which arised directly from the egg cell of the unreduced embryo sac, and the frequency of polyembryonic seed was 34.25% and 8.11% respectively. The number of embryos per embryonic seed ranged from 2 to 3. The frequency of multiple seedlings in *Poa pratensis* was 6.14%. Based on the above observations Cai *et al.* (1997) concluded that the phenomena of polyembryony and multiple seedlings of apomictic plant could be used as sketchy screening indicator for identifying plants tending to be apomictic.

Tsvetova and Ishin (1996) reported that in *Sorghum* as the sexual embryonic sac is usually in the seed-bud in constant position and the sexual embryo has the definite position in the embryo sac, it has been supposed that additional embryo unusual position in the seed indicates is apomictic origin.

CHAPTER 10

AN EMBRYOLOGICAL APPROACH TO THE TAXONOMY AND PHYLOGENY OF POACEAE

Most systems of classification of plants have been evolved on the basis of exomorphic characters placing particular reliance on the characters of the flower, the flower being considered more conservative than the vegetative organs. Even so the goal of reaching a truly natural system of classification has remained unfulfilled. The importance of embryological data as an additional evidence to that obtained from the fields of cytology, anatomy, palynology, ultrastructure and phytochemistry in elucidating the taxonomic relationships is well known and it has been emphasised by eminent embryologists e.g., Schnarf (1933, 1937), Just (1946), Johansen (1945, 1950), Maheshwari (1950, 1963, 1964), Lebegue (1952), Cave (1953, 1959), Johri (1963), Poddubnaja-Arnoldi (1964, 1976), Palser (1975) and Herr (1984). Herr (1984) listed a number of embryological features of taxonomic significance which are taken into consideration for evaluating phylogenetic relationships.

Embryological evidence must be weighed against other criteria, e.g., cytological, anatomical, palynological, ultrastructural and phytochemical evidence. The systematic position and phylogeny of the family Poaceae presented in this chapter is based on the embryological data supplemented by other kinds of information known to have systematic and phylogenetic value.

The Systematic Position of the Family Poaceae

Diverse opinions have been expressed regarding the systematic position of the family Poaceae in various systems of classification by different taxonomists. Lindley (1853) placed Gramineae and Cyperaceae under the order Glumales in the class Endogenes, which also include Devauxiaceae (Centrolepidaceae), Restionaceae and Eriocaulaceae. Bentham and Hooker (1862–1883) included Gramineae in the series Glumaceae along with the families Eriocaulaceae, Centrolepidaceae, Restionaceae and Cyperaceae. Van Tieghem (1891) divided the Monocotyledons into four orders and included the families Gramineae,

Cyperaceae and Centrolepidaceae in order Graminees. Hallier's (1912) order of Enentioblastae include families Eriocaulaceae. Centrolepidaceae, Restionaceae, Flagellariaceae, Rapataceae, Xyridaceae, Philydraceae, Pontederiaceae, Commelinaceae and Mayaceae. Hallier separated the family Cyperaceae and placed it in the order Cyperales along with Juncaceae. Rendle (1930, 1938) placed Gramineae after the Triuridales due to the wind pollinated, perianthless flowers. Engler (1887) included the Gramineae and Cyperaceae under the order Glumiflorae. Warming (1904), Lotsy (1911) also followed the same relative positions for these families. Engler and Prantl (1897–1915) order Glumiflorae has been considered by many botanists to compose a natural assemblage, and one with which some phylogenists in the past included other families like Juncaceae, Centrolepidaceae, Thurniaceae and Restionaceae. Engler and Diels (1936), Bensen (1957) also placed Gramineae in the order Graminales along with Cyperaceae. However the placing of both Cyperaceae and Gramineae into one order was rejected by Hutchinson (1934, 1948, 1969) who separated the two families and raised them into two different orders Cyperales and Graminales. He placed these two orders along with a third order Juncales in the Division Glumiflorae. The results of morphological studies by Belk (1939) and by Balser (1940, 1944) also supported the segregation of Cyperaceae and Gramineae. But they failed to provide substantiating evidence. Takhatajan (1969) separated Cyperaceae and Gramineae and kept them under separate orders Cyperales and Graminales based on the characters of the stem, leafsheath, arrangement of the flower, bracts, pollen grains, nature of the ovule, embryo and kind of seed etc. Melchoir's (1964) order Graminales is essentially similar to that of Hutchinson (1969) and Takhtajan (1980) in comprising the only family Gramineae. Bessey's (1915) order Gramineae is simiar to that of Bentham and Hooker (1862–1883) excepting that he separated the family Eriocaulaceae and included the family Flagellariaceae in the order Graminales. He placed the family Eriocaulaceae in the order Liliales along with Liliaceae, Stemonaceae, Pontederiaces, Cyanastraceae, Philydraceae, Commelinaceae, Xyridaceae, Mayaceae, Juncaceae, Thurniaceae, Rapataceae and Naiadaceae. Thorne (1968) included the families of Bentham and Hooker (1867-1883) and Hallier (1912) in the order Commelinales. However, Thorne's (1968) order Commelinales differs from Hallier's (1912) order Enentioblastase in that he separated the family Cyperaceae and kept it in the order Cyperales along with Juncaceae. Dahlgren (1975) included the family Poaceae in the order Poales along with Restionaceae, Flagellariaceae, Ecdeiocoleaceae and Joinvillieaceae. Cronquist (1968, 1981) included Cyperaceae and Gramineae in his order Cyperales.

Table V gives the systematic position of the family Poaceae in the various systems of classification. For convenience the morphological features of different families are also tabulated in the last column of the table. The embryological features of the family Poaceae and those of the other families included in the orders Glumiflorae/Graminales/Commelinales are tabulated in Table VI.

Table V. Poaceae in different systems of classification

Bentham & Hooker (1862-1883)	Hallier (1912)	Engler & Prantl (1897-1917) Engler & Diels (1936)	Bessey (1915)	Benson (1957)	Melchoir (1964)	Thorne (1968)	Hutchinson (1969)	Dahlgren (1975)	Takhtajan (1980)	Cronquist (1981)	Characters of the families
Glu-maceae	Enentio-blastae	Glumi-florae	Grami-nales	Grami-nales	Grami-nales	Comme-linales	Grami-nales	Poales	Poales	Cyperales	
						Bromel-iaceae					Mostly epiphytes. Stems reduced with rosette of fleshy leaves. Inflorescence usually from the centre of the pitcher. Flowers bisexual, regular trimerous, perianth in two whorls, free or fused. Stamens 6.Ovary tricarpellary, syncarpous inferior, semi-inferior or superior with numerous ovules on axile placentation. Fruit berry or capsule.
	May-aceae					May-aceae					Aquatic herbs with alternate, simple entire, non-sheathing leaves. Shortly bifid at

(contd.)

(Table V contd.)

Bentham & Hooker (1862-1883)	Hallier (1912)	Engler & Prantl (1897-1917) Engler & Diels (1936)	Bessey (1915)	Benson (1957)	Melchoir (1964)	Thorne (1968)	Hutchinson (1969)	Dahlgren (1975)	Takhtajan (1980)	Cronquist (1981)	Characters of the families
											apex. Flowers regular, bisexual, sepals three, petals three, stamens three, ovary superior, tricarpellary, syncarpous, unilocular with many biseriate ovules on parietal placentation. Fruit 3-valved capsule.
	Xyrid-aceae					Xryid-aceae					Mostly marsh plants, herbaceous, inflorescence spike or head of bisexual flowers. Perianth heterochlamydeous. Sepals 3, petals 3, stamens epipetalous, outer whorl absent or staminode, ovary tricapellary, syncarpous, superior, locules 1-3, many ovules with parietal or free basal placentation. Fruit capsule.
	Rapat-aceae					Rapat-aceae					Perennial herbs from fleshy root stocks. Inflorescence terminal head of spikelets enclosed in two large spathes.

		Each spikelet with numerous bracts and terminal one floret. Sepals three, petals three connate at base. Stamens 6. Ovary superior. Ovules many, arranged in axile or basal placentation. Fruit loculicidal capsule.
	Ecdio-cole-aceae	Glabrousperennial herbs with simple erect slender cylindrical stem from a creeping rhizome. Inflorescence terminal, conical or cylindrical spike. Flowers unisexual, monoecious, perianth in two whorls of three each, unequal, glumaceous, stamens 3-4, ovary superior, bicarpellary, one ovule per locule.
Ponte-deri-aceae	Ponte-deri-aceae	Water plants, floating or rooted. The successive ending in inflorescence (Sympodial cymose, pseudo racemes).Flowers zygomorphic, perianth in two whorls, fused and persistent. Stamens 6 or 2-3, epipetalous. Ovary superior,

(contd.)

(Table V contd.)

Bentham & Hooker (1862-1883)	Hallier (1912)	Engler & Prantl (1897-1917) Engler & Diels (1936)	Bessey (1915)	Benson (1957)	Melchoir (1964)	Thorne (1968)	Hutchinson (1969)	Dahlgren (1975)	Takhtajan (1980)	Cronquist (1981)	Characters of the families
											locules 3 with many ovules or unilocular with single ovule. Fruit capsule or nut.
	Comme-linaceae						Comme-linaceae				Usually herbs (occasionally twining) with jointed stems. Inflorescence usually cincinnus like Boraginaceae. Flower bisexual, sepals 3, petals 3, rarely fused. Stamens 6, some often absent or staminode. Ovary superior. Locules 3. Fruit loculicidal capsule. Seed often with aril; rarely winged.
	Philydr aceae						Philydr-aceae				Erect herbs with 2 ranked sheathing narrow leaves. Inflorescence simple or compound spikes. Flowers bisexual, zygomorphic, perianth homochlamydeous, stamen one, ovary superior, ovules many, with axile or parietal placentation. Fruit capsule.

			Junc-aceae			Usually creeping sympodial rhizome, one part of the sympodium appear above the ground as a leafy shoot, stem does not lengthen above ground except to bear inflorescence which is crowded mass of flowers in cymes. Flowers bisexual, perianth in two whorls of 3 each, sepaloid, stamens 6. Superior ovary, ovules many or few on axile placentation or parietal placentation. Fruit loculicidal capsule.
Eriocaulaceae	Erio-caula-ceae			Erio-caul-aceae		Perennial herbs with often grass-like leaves. Flowers in involucral heads, unisexual, 2-3-merous, regular or zygomorphic. Perianth in usually whorls, male flowers with stamens 4-6 or 3-2 with two or one-thecous anthers. Female flowers with superior ovary, 1-3-locular. Fruit loculicidal capsule.
Cyperaceae		Cyperaceae	Cyper-aceae	Cyper-aceae	Cyper-aceae	Leaves 1/3 hyllotaxy, ligule absent. Inflorescences one to many flowered spike-like cymes. Flowers born in axils of only one glume; bracteoles and perianth usually absent.

(contd.)

(Table V contd.)

Bentham & Hooker (1862-1883)	Hallier (1912)	Engler & Prantl (1897-1917) Engler & Diels (1936)	Bessey (1915)	Benson (1957)	Melchoir (1964)	Thorne (1968)	Hutchinson (1969)	Dahlgren (1975)	Takhtajan (1980)	Cronquist (1981)	Characters of the families
											Tepals if present either six in two whrols or reduced to hair and bristles or sometimes three. Ovary tri- or bi-carpellary, syncarpous, superior and unilocular with single basal ovule. Fruit achene.
	Flagell-ariaceae		Flagell-ariaceae			Flagell-ariaceae		Flagell-ariaceae			High climbing lianas from diffuse sympodial rhizome. Stems frequently branched by equal dichotomy, internodes solid, covered by leaf sheaths. Leaf sheath closed. Secretory cells present. Inflorescence terminal, branched, bracteate. Flowers bisexual perianth in two whorls of three each. Stamens six, ovary superior, tri-carpellary with one ovule per locule on axile placentation. Fruit drupaceous.

	Centrolepi-diaceae	Centrolepi-diaceae		Centro-lepidi-aceae	Centrolepi-diaceae		Small annual or perennial grass-like herbs. Flowers small, bisexual or unisexual, naked or surrounded by one to three trichous bracts. Perianth absent. Stamens one to two. Ovary superior, unilocular with single ovule.
Restion-aceae	Restion-aceae		Restion-aceae		Restion-aceae	Restion-aceae	Xerophytes, usually of tufted growth. Shoots bearing sheathing leaves which have short blade or sometimes none. Flowers dioecious. Inflore-scence spikelets. Perianth in two whorls. Stamens three or two opposite to the inner perianth. Ovary superior, tricarpellary or with single carpel, tri-or bilocular. Fruit capsule or nut.
						Joinville-aceae	Erect herbs. Stems unbranched, internodes hollow. Leaf dorsiventral, leaf sheath open. Secretory cells absent. Inflorescence terminal, much branched, bracteate. Flower regular bisexual, sessile. Perianth in two whorls of 3 each, dry, bract-like, persistent.

(contd.)

(Table V contd.)

Bentham & Hooker (1862-1883)	Hallier (1912)	Engler & Prantl (1897-1917) Engler & Diels (1936)	Bessey (1915)	Benson (1957)	Melchoir (1964)	Thorne (1968)	Hutchinson (1969)	Dahlgren (1975)	Takhtajan (1980)	Cronquist (1981)	Characters of the families
Gramineae	Gramineae	Gramineae	Poaceeae	Gramineae	Gramineae	Gramineae	Gramineae	Poacnae	Poaceae	Poaceae	Stamens six, ovary superior, tricarpellary. Single ovule per locule. Fruit drupaceous. Tufted or stoloniferous herbs. Leaves alternate, usually narrow, ligule present. Inflorescence compound spike, flowers bisexual or unisexual, zygomorphic. Perianth usually represented by two minute hypogymous scaly lodicules. Stamens usually three. Fruit caryopsis.

Table VI. Embryological features of Poaceae and related families

Character	Bromeliaceae	Ponte-deriaceae	Commelin-aceae	Philydraceae	Juncaceae	Eriocaulaceae	Cyperaceae	Flatgellari-aceae	Centrolepid-aceae	Restion-aceae	Poaceae
Anther	—	Tetra-sporangiate	Tetra-sporangiate	Bispo-rangiate	Tetra-sporangiate	Tetra-sporangiate	Tetra-sporangiate	—	Bispora-ngiate	Bispora-ngiate	Tetra-sporangiate
Male archesporium	—	—	—	—	—	—	Single-layered	—	—	—	Single-layered
Anther wall development	—	—	—	—	—	—	Monocotyle-donous type	—	Monocotyle donous type	— —	Monocotyle-donous type
Anther tapetum	Glandular	Amoeboid, cells 2-nucleate	Amoeboid,	Glandular, cells 2-nucleate	Glandular, cells 2-nucleate	—	Glandular, cells 2-nucleate	—	—	Glandular, cells 1-nucleate	Glandular, cells 1-nucleate, rarely 1-nucleate
Pollen tetrads	—	Isobilateral or decussate	Isobilateral or decussate	Isobilateral	Tetrahedral or Isobilateral	—	Tetrahedral decussate or isobila-teral	Isobilateral or linear	Linear, T-shaped,	—	Isobilateral rarely T-shaped, linear or decussate
Pollen grains	2-celled	2-celled	2-celled or 3-celled	2-celled	2-celled	3-celled	3-celled	—	3-celled	3-celled	3-celled
Ovules	Antaropous, bitegmic and crassi-nucellate	Antaropous, bitegmic and crassi-	Orthotropous-hemiana tropous, bitegmic and	Antaropous, bitegmic and crassi-nucellate	Antaropous, bitegmic and crassi-nucellate	Orthotropous, bitegmic and tenuinucellate	Anatropous, bitegmic and crassinucellate	Orthotropous	Orthotro-pous, bitegmic and	Orthotro-pous, bitegmic and	Anatropous, hemiana-tropous or campylo-

(contd.)

(Table VI contd.)

Character	Bromeliaceae	Pontederiaceae	Commelinaceae	Philydraceae	Juncaceae	Eriocaulaceae	Cyperaceae	Flatgellariaceae	Centrolepidaceae	Restionaceae	Poaceae
											tropous, bitegmic, tenui-nucellate or pseudocrassi-nucellate
Parietal tissue	Present, 6 layers	Present, only one layer	Present, only one layer	Present, only one layer	Present, 1-2 layers	Absent	Present, 4 layers	—	Absent	Absent, rarely 1 layer present	Present, 1-6 layers in Panicoideae. Absent in Pooideae
Embryo sac	Polygonum type	Polygonum type	Polygonum type, rarely Adoxa type	Polygonum type	Polygonum type	Polygonum type	Polygonum type	Allium type	Polygonum type	Polygonum type	Polygonum type
Antipodals	Three, ephemeral	Three, ephemeral	Three, ephemeral	Three, ephermeral	Three, persistent upto early embryo	Three, ephemeral	Three, ephemeral	Three, ephemeral	Three, persist up up to fertilization	Three, ephemeral, rarely upto 15	Three to three hundred
Endosperm	Helobial	Helobial	Nuclear	Helobial	Helobial	Nuclear or Helobial	Nuclear	—	Nuclear	Nuclear	Nuclear
Endosperm haustrorium	—	—	Present	—	—	—	Absent	—	—	—	Absent
Embryo	Asterad type	Asterad type	Asterad type	Onagrad type	Onagrad type	Asterade type	Onagrad type	—	Onagrad type	—	Asterad type

The families Eriocaulaceae, Flagellariaceae, Rapataceae, Xyridaceae, Philydraceae, Pontederiaceae, Commelinaceae and Mayaceae of Hallier (1912) and Thorne (1968) are included along with the family Gramineae. In addition to these above families Bromeliaceae and Juncaceae are also treated along with Gramineae by Thorne (1968).

The family Eriocaulaceae is treated along with Mayaceae, Xyridaceae, Restionaceae and Centrolepidiaceae by Engler and Diels (1936), Eriocaulaceae shows some similar embryological features with Gramineae such as presence of tetrasporangiate anthers, 3-celled pollen grains, Polygonum type of embryo sac, and Asterad type of embryogeny and differs from it by possessing orthotropous ovules, lack of periclinal divisions in the nucellar epidermis, presence of three ephemeral antipodals, Nuclear or Helobial type of endosperm development.

Hutchinson considers the family Eriocaulaceae as sufficiently apart from the other families to be elevated to a separate order Eriocaulales, and advanced over his Xyridales on the basis of flowers always unisexual and ovules reduced to one per carpel.

The families Bromeliaceae, Philydraceae, Pontederiaceae, Commelinaceae differ from Gramineae in some embryological features. In Pontederiaceae bisporangiate anthers are reported whereas in Gramineae anthers are tetrasporangiate. In Pontederiaceae and Commelinaceae anther tapetum is of Amoeboid type whereas in Gramineae it is always Glandular. The pollen grains are 3-celled in Gramineae, but in Bromeliaceae, Philydraceae, Pontederiaceae and Commelinaceae they are 2-celled. The ovules in Gramineae are anatropous, hemianatropous or campylotropous whereas in Bromeliaceae, Philydraceae, Pontederiaceae they are anatropous while in Commelinaceae they are orthotropous. The antipodals are usually three and ephemeral in all these families whereas in Gramineae they proliferate. The development of endosperm in Bromeliaceae, Philydraceae and Pontederiaceae is Helobial type and that of Gramineae is of Nuclear type.

By the above embryological evidences, the family Gramineae must be separated from the above families and should be regarded as a separate order as treated by Melchoir (1964), Hutchinson (1969) and Takhtajan (1980).

Hutchinson (1934) treated the family Bromeliaceae as related to Commelinales but more advanced from it. Smith (1934) considered the strongest affinities of the family with Rapataceae.

Based on the endosperm and embryo character the family Pontederiaceae was included in the order Farinosae of Engler and Prantl. Hutchinson placed Pontederiaceae in the order Liliales stating that "they appear to me to be aquatic Liliaceae, tending towards the Aroid type, the spiciform inflorescence having spathe like reduced leaf (leafsheath)." Schwartz (1930) also treated the family as being closely related to the Liliaceae, but considered the characters of endosperm, the variability and reduction in the androecium and the floral zygomorphy to be of sufficient importance to justify placing it close to Commelinaceae.

The family Flagellariaceae is included in the order Poales/ Enentioblastae/ Graminales/Commelinales along with Gramineae by Hallier (1912), Bessey (1915), Thorne (1968) and Dahlgren (1975). Embryologically the family Flagellariaceaeae is inadequately known. Dahlgren (1981, 1983) remarked that the family Flagellariaceae deserves a place in the order Poales since it has several features in common with grasses. Takhtajan (1969) is also of the opinion that the Flagellariace of the Restionales is most similar to the Gramineae.

Thorne (1968) treated the family Juncaceae along with Gramineae. It shows common embryological features with grasses in possesing tetrasporangiate anthers, Glandular anther tapetum, 3-celled pollen grains, presence of parietal layers in the micropylar region, Polygonum type of embryo sac. It differs from Gramineae by simultaneous cytokinesis, anatropous, crassinucellate ovules, three antipodals, Helobial type of endosperm development and Onagrad type of embryo development. Engler and Diels (1936) included it in the suborder Juncineae of the order Liliflorae. Hutchinson (1959, 1969) placed it in an order Juncales treating it as advanced over his Liliales. He treated it as one of the three orders comprising his class Glumiflorae and as a progenitor of his Cyperales and Graminales.

The two families Xyridaceae and Mayaceae are also treated along with Gramineae by Hallier (1912) and Thorne (1968). Embryologically these families are inadequately known. Engler and Prantl (1897-1915) considered that Mayaçeae is perhaps closer to Flagellariaceae and Commelinaceae and the family Xyridaceae as a separate order Xyridales and considered them to represent by reduction, an advancement over Commelinales.

The Joinvilleaceae of Dahlgren (1975) have many features in common with grasses (Dahlgren and Cufford 1982) but the development of the embryo sac has not been described as to allow considerations on their possible embryological affinities. At present situation, the embryological exploration of the Joinvilleaceae ranks first to set fundamental information for the comprehension of its possible phylogentic relationships with the Poaceae (Anton and Cocucci, 1984).

An analysis of data from Table V and VI clearly shows that the family Cyperaceae differs from the family Poaceae in its morphological as well as embryological features. Recent investigations based on the broad comparative morphology and evidences from different disciplines of botany do not favour a close relationship between the two families.

Morphologically Cyperaceae is characterised by solid and often triquetrous stems, 1/3 phyllotaxy with the leaf sheath generally closed, absence of ligule and lodicules, position of meristems in the leaf, nature and structure of spikelets and glumes, absence of perianth and the type of fruit, while the family Poaceae is characterised by the presence of stems with nodes and internodes, alternate leaves, presence of ligule, presence of intercalary meristem, compound spikelike inflorescence, presence of parinath, type of fruit i.e., the caryopsis with testa fused with the pericarp.

The embryological investigations also reveals various differences between the two families Cyperaceae and Poaceae. In Cyperaceae cytokinesis is of simultaneous type whereas in Poaceae it is of successive type. In Cyperaceae only one microspore in the tetrad is functional and the remaining three microspores degenerate whereas in Poaceae all the four microspores of a tetrad are functional. The antipodals are 3, ephemeral in Cyperaceae while they form a complex or become coenocytic in Poaceae. The embryogeny is of Onagrad type with Juncus variation in Cyperaceae whereas in Poaceae it is Poa variation of Asterad type. The testa and pericarp are distinctly free in Cyperaceae and the testa comprises both the integuments. In Poaceae testa and pericarp are fused and either both the integuments are obliterated or only the inner integument forms the testa.

In addition to the above differences the two families differ from each other in the nature of thickenings developed on endothecial cells and associated globular markings or ubisch granules, character of germ pore, development, structure and form of ovule, structure and organisation of micropyle, behaviour of nucellar epidermis, presence or absence of nucellar cap, nature and behaviour of female archesporium, structure of synergids and presence or absence of associated filiform apparatus, number, position and nature of antipodals and their persistence in the developing seed, presence or absence or hypostase, nature of mature endosperm and its superficial layer development and structure of mature embryo, occurrence or absence of apomixis.

Inspite of these differences Cronquist (1968; 1981, 1988) sparked off controversy by strongly advocating relationship between these two families. He placed these two families in a single order Cyperales. The Cyperales of Cronquist (1981) is one of the 7 orders under the sub-class III Commelinidae of the class Liliopsida. He believes that it is more useful to group Gramineae and Cyperaceae in a single order Cyperales. Based on chemical data Cronquist (1968) says that "although the Cyperaceae differs from the Gramineae in several respects, the chemical similarities between them when considered in the context of more classical characters, are more suggestive of phyletic unity than convergence".

Hegnauer (1963) finds that the admittedly scanty chemical data fully compatible with a relationship among Juncaceae, Cyperaceae, Restionaceae and Gramineae. Furthermore although the Cyperaceae differs from the Gramineae in several respects, the chemical similarities between them are more suggestive of phyletic unity than convergence. In recent years many authors have segregated Cyperaceae and Gramineae into different orders and even suborders. The view has been conditioned partly by the pseudanthial interpretation of the Cyperaceous flower and partly by the seemingly incompatible relationships of the two families Cyperaceae to the Juncale and Gramineae to the Restionales. However many

Agrostologists recognised considerable difference between Cyperaceae and Poaceae.

The detailed morphological and anatomical investigations by Snell (1936) and Blaser (1940; 1941a,b, 1944) indicated that the Gramineae are not close allies of the Cyperaceae. It was pointed out by Blaser (1940, 1944) that (i) superficial grass like structure of Cyperaceae is not of much phylogenic significance and has appeared also in other non related families, (ii) the spikelets of Cyperaceae are not at all homologous with those of grasses; (iii) the basic placental condition of Cyperaceae has been derived from an ancestral free central type whereas in Gramineae it has been derived from a parietal type, (iv) florests of Gramineae are born in terminals whereas in the Cyperaceae they are always axillary in position. It is clear from these considerations that these families belong to two separate orders.

It is clearly evident from all the above observations that the two families Cyperaceae and Poaceae show remarkable differences in morphological and embryological characters and it is not favourable to retain the family Cyperaceae along with Poaceae as treated by Bentham and Hooker, Engler and Diels, Engler and Prantl, Bensen, Thorne and Cronquist. It is therefore proper to treat the two families separately under two separate orders Cyperales and Graminales/ Poales as treated by Melchoir (1964), Hutchinson (1969) and Takhtajan (1980)

Inspite of several morphological, anatomical and embryolgoical differences, many similarilies have been noticed between the embryological features of the Cyperaceae and Poaceae. The common characters are the presence of (a) usually tetrasporangiate anthers, (b) monocotyledonous type of anther wall development, (c) Glandular and 2-nucleate tapetal cells, (d) 3-celled pollen grains, (e) micropyle formed by the inner integument, (f) Polygonum type of embryo sac and (g) Nuclear endosperm. From these characters it is clear that the family Cyperaceae is the nearest to the family Poaceae and should be placed in the immedately preceding order.

The families Centrolepidaceae and Restionaceae are included along with Gramineae by Bentham and Hooker (1862-1883), Hallier (1912), Bessey (1915) and Thorne (1960). Embryologically these two families show close similarity like presence of usually (a) bisporangiate anthers with persistent epidermis, (b) secretary anther tapetum, orthotropus, bitegmic and tenuinucellate ovules, (c) microphyle formed by both integuments, (d) Polygonum type of embryo sac, (e) starch grains in the embryo sac, (g) Nuclear endosperm, (f) radial elongation of apical cells of nucellar epidermis and (h) fusion of polar nuclei before fertilization.

The Gramineae on the other hand, while showing an overall resemblance in features like: Monocotyledonous type of anther wall development, presence of secretary anther tapetum, bitegmic ovules, Polygonum type of embryo sac development, Nuclear type of endosperm development and fusion of polar nuclei before fertilization differ in exhibiting (a) tetrasporangiate anther, (b) anatropous to campylotropous ovules, (c) a micropyle usually formed by inner integrument, (d) periclinal divisions in nucellar epidermal cells to form a cap of up to 6 layers (except in Pooideae), (e) meristematic activity in peripheral layers of endosperm and (f) Asterad type of embryogeny and common occurrence of polyembryony. Hence the Gramineae do not seem to be as close, from the point of view, to Centrolepidaceae and Restionaceae. These two families can be regarded as one of the closest families to the Poaceae.

Dahlgren (1975) considered the family Restionaceae to be the most related to Poaceae as both the families exhibit in common many sporophytic characters. Additionally, they also show a concidence at the embryo sac organization, since they have the same type of megagametophyte in which antipodals proliferate. However embryological characters show several differences between these two families.

SUB-DIVISION IN THE FAMILY POACEAE

In most of the systems of classification the family Poaceae is divided into two sub-families. Brown (1814) for the first time sub-divided the grasses into sub-families Pooideae and Panicoideae. Haeckel (1887) also divided the family into two sub-families as done by Brown, but he named Pooideae as Festucoideae. Lawrence (1951) also stated that the genera of the family Poaceae are generally grouped by agrostologists under the two sub-families Festucoideae and Panicoideae.

An analysis of the embryological features of this family reveals that the two sub-families Pooideae and Panicoideae exhibit striking differences, especially those relating to the ovule, the development of outer integument, presence or absence of parietal layers in the micropylar region, type of ovule, orientation of the organised embryo sac, number and position of antipodals and the embryo. Details of the structure of the embryo are also considered to be of critical importance in the definition of sub-families of grasses. Reeder (1957, 1962) proposed a formula to denote embryo types based on four distinctive features, the course of vascular bundles in the embryo, the presence or absence of an epiblast, the presence or absence of a groove between the lower part of the scutellum and the coleorhiza and the conduplicate or convolute vernation of the primordium of the first leaf of the plumule.

Different embryologists like Maze *et al.* (1970; 1991), Chandra (1963 a,b), Bhanwra (1988), Narayanaswami (1953; 1955a,b,c; 1956) and Diwanji (1976) have reported that the ovules of Pooideae and Panicoideae differ considerably in the mode of development of outer integument, presence or absence of periclinal divisions in the nucellar epidermis, presence or absence of enlarged nucellar cells below the micropyle and the position of antipodals in the embryo sac after fertilization.

Panicoideae differs from the sub-family Pooideae in possessing poorly developed outer integument, parietal layers in the micropylar region, chalazal position of antipodals, embryo sac oriented almost parallel to the long axis of the ovule, Pancoid type of embryo, absence of epiblast and persistence of nucellar epidermis in mature grain. On the other hand Pooideae is characterised by well developed inner and outer integuments, lack of parietal layer in the micropylar region, lateral position of antipodals in the embryo sac after fertilization, Festucoid type of embryo, presence of epiblast, obliterated nuclear epidermis and development of seed coat from the inner intigument (Table VII).

Reeder (1957) made an extensive study on the embryos of 300 species belogning to 150 genera and his observations on the structure of mature embryo clearly show distinct differences between the two sub-families Panicoideae and Pooideae. In Panicoid embryos the traces to the scutellum and embryonic leaves are separated by an elongated internode, epiblast is abesent, lower part of the scutellum is fused with coleorhiza and embryonic leaves show few bundles and their margins are merely meet.

In Festucoid embryos the traces to the scutellum and embryonic leaves diverge approximately at the same point, epiblast is present, lower part of the scutellum is free from the coleorhiza and embryonic leaves show many vascular bundles and overlapping margins.

The microcharacters that are used in the systematics e.g., shape of silicified epidermal cells and structure of starch grains are often constant within whole tribe but are of secondary importance in characterizing taxa a rank as sub-family. In Panicoid grasses the silicified cells are commonly dumb-bell shaped whereas in Festucoid grasses the silicified cells are not highly variable and are usually simpler. Bicellular microhairs are never found in Festucoid grasses, whereas in the other group they are usually present (Tzvelev, 1989).

On the basis of microstructural characters, all grasses were divided in to six main groups: bambusoid grasses,festucold grasses, oryzoid grasses, arundinoid grasses, eragrostoid grasses (chloroid grasses) and panicoid grasses. These groups become widely accepted in the literature as sub-families (Tzvelev, 1989).

Several other authors added several sub-families to provide either groups of genera intermediate between other families, or for genera with very distinctive characteristics. Caro (1982) divided the family Gramineae into 13 sub-families namely Bambusoideae, Streptochaetoideae, Anomochlocoideae, Glyroideae, Centrostecoideae, Oryzoideae, Ehrhartoideae, Phragmitoideae, Festucoideae, Eragrostideae, Aristidoideae, Panicoideae and Micrairoideae.

Tzvelev (1989) considered that taxa of sub-familial rank differ from taxa of tribal rank in that they may be segregated as independent families. But this cannot be said about most of the proposed sub-families of grasses. In this respect sub-families are comparable to sub-genera which can be accepted only for those sections which might also be recognised at generic rank (Tzvelev, 1989).

Table VII. Contrasting embryological features of the two sub-families Pooideae and Panicoideae

Panicoideae	Pooideae
1. Ovoid ovary	1. Obovate ovary
2. Dorsal ovary wall of the same thickness in the distal region and near the base.	2. Dorsal ovary wall thicker in the distal region and narrowing towards the base.
3. Outer integument is poorly developed and enclosing only about one third of the ovule.	3. Both the integuments are well developed.
4. 2-7 periclinal divisions in nucellar epidermis in the vicinity of the micropyle is of regular occurrence.	4. Periclinal divisions in the nucellar epidermis in the vicinity of the micropyle are absent or very rarely present.
5. Presence of enlarged nucellar or parietal cells in many species.	5. Absence of enlarged nucellar cells below the micropyle
6. Mature embryo sac oriented to the longitudinal axis of the ovule.	6. Mature embryo sac showing curvature towards the chalaza.
7. Embryo sac ovate or spindle-shaped.	7. Embryo sac obovate or rarely oblong during free nuclear stages of endosperm.
8. No change in the position of antipodals.	8. Displacement of antipondals to a lateral position in the embryo sac after fertilization.
9. No deposition.	9. Deposition of darkly staining material in the inner epidermis of the inner integument after ferilization.
10. The apomictic (unreduced) embryo sacs are usually monopolar and 4-nucleate.	10. The apomictic (unreduced) embryo sacs are usually bipolar and 8-nucleate.
11. Embryo:	11. Embryo:
i. Panicoid type.	i. Festucoid type.
ii. The traces are separated by an elongated internode.	ii. The vascular traces to the scutellum and embryonic leaves diverge at the same point.
iii. Epiblast is absent.	iii. Epiblast is present.
iv. Lower part of the scutellum is fused with coleorhiza.	iv. Lower part of the scutellum is free from the coleorhiza.
v. Embryonic leaves show many bundles.	v. Embryonic leaves show few bundles.
12. Persistent nucellar epidernis in the mature grain.	12. Obliterated nucellar epidemis in the mature grain.
13. Aleurone transfer cells conspicuous.	13. Aleurone transfer cells inconspicuous.

Tzvelev (1989) recognised only two sub-families of grasses - Bambusoideae and Pooideae. His classification is based on the characters like type of stem, presence or absence of true petiole and structure of embryo. According to him: (i) the stems (culms) of the bamboo grasses are woody and often branched upward, whereas Pooid grasses are almost herbaceous, (ii) the leaves of bambusoid grasses have true petiole whereas among the Pooid grasses leaves with poorly developed petioles are found only in few tropical genera, (iii) in the pooid grasses a meristematic layer is formed under the base of the plumule.

Through its activity the base of the plumule can lengthen considerably beneath the coleoptile during germination and adventitious roots can form beneath the coleoptilar node, whereas bamboos do not have such meristem and as a result coleoptile base remains in place during germination and adventitious roots do not develop at or below the coleoptilar node. The proposed system of Tzvelev (1989) is as follows.

i) Sub-family Bambusoideae

Consisting of 14 tribes namely Arundinarieae, Shibataceae, Bambuseae, Dendrocalameae, Bacciferae, Oxytenanthereae, Atractocarpeae, Streptogyneae, Steptochaeteae, Buergersipchloeae, Olyreae, Parianeae, Leptaspideae, Anomochloeae.

ii) Sub-family Pooideae

Consists of 27 tribes such as Brachypodieae, Triticeae, Bromeae, Poeae, Phleeae, Meliceae, Brylkinieae, Diarrheneae, Brachyeltytreae, Ampelodesmeae, Stipeae, Lygeeae, Nardeae, Phaenospermateae, Oryzeae, Phyllorachideae, Ehrharteae, Centosteceae, Arundineae, Thysanolaeneae, Micraireae, Aristideae, Cynodonteae, Arundinelleae, Isachneae, Paniceae and Andropogoneae.

The sub-family Bambusoideae according to Tzvelev (1989) comprises 111 genera. Embryologically this sub-family is little investigated. Out of 111 genera only 5 genera *Bambusa, Dendrocalamus, Sasa, Thyrsostachyus and Melocanna* have been studied embryologically. Unless otherwise embryological studies are carried out on other genera the raising of Bambusoideae into a sub-family cannot be discussed on embryological grounds. Grouping of Panicoid and Pooid grasses into one sub-family Pooideae by Tzvelev (1989) is not justified on embryological grounds as these two sub-families show clear cut differences in morphological characters (See Table VII).

ORIGIN AND PHYLOGENY OF POACEAE

The Gramineae is a very ancient family which evidently originated not later than the Cretaceous period. From the beginning of the Tertiary period fairly numerous and sufficiently reliable remains of grass-like leaves are found on many continents. Morphologically and taxonomically the grasses have been subjected to various and divergent opinions.

Different views exist regarding the origin of the family Poaceae. According to one school of taxonomists the family has been evolved from the primitive Liliaceous ancestral stocks. Some other taxonomists are of the opinion that Palms (Arecaceae) are the possible ancestral stock to the family Poaceae. Cronquist (1968) is of the opinion that the four families Juncaceae, Cyperaceae,

Restionaceae and Gramineae are evolved from Commelinales. Dahlgran (1980) is also of the opinion that the family Poaceae have been evolved from Commeliniflorae. Takhtajan (1980), Cronquist (1988), Dahlgren (1983), Dahlgren and Clifford (1982) considered the members of Restionales as possible ancestors of Poaceae. Brewbaker (1967) viewed that the order Poales is derived from the order Juncale through which the order Cyperales has also been derived. Hutchinson (1959, 1973) is also of the opinion that the order Juncales is progenitor of his Cyperales and Poales.

I. Arecaceous Origin

According to one group of taxonomists the family Poaceae has been derived from Arecaceae (Palms). Clifford (1970) speculated the possible evolution of grasses from the palms. He investigated inter-relationships of some monocot families with special reference to the origin of grasses. However the family Arecaceae differs from Poaceae in several characters such as woody habit, stems covered by persistent leaf bases, large compound leaves, compound spadix inflorescence, usually unisexual flowers, perianth in two whorls which is of woody nature, tricarpellary, apocarpous or partially connate or syncarpous gynoecium, trilocular ovary, anatropous ovules and fruit being either drupe or berry.

In possessing the following embryological characters the family Arecaceae differs from Poaceae. The characters are: 2-6 ephemeral middle layers of anther wall, irregularly arranged 2-layered glandular tapetum, simultaneous cytokinesis, 2-celled pollen grains, anatropous to hemianatropous, crassinuallate ovules, micropyle formed by both the integuments, three ephemeral antipodals and Onagrad type of embryogeny (Davis, 1966).

From the above observations it is very clear that the two families are not similar in several morphological as well as embryological features. Hence the opinion of Palm ancestory to the grasses as suggested by Clifford (1970) cannot be accepted and it is untenable.

II. Commelinales Origin

Cronquist (1968) believes that the four related families Juncaceae, Cyperaceae, Restionaceae and Gramineae are evolved from Commelinales. Dahlgren (1980) also proposed that the order Poales has been derived from Commeliniflorae along with the orders Eriocaulales, Cyperales, Juncales and Commelinales. Cronquist considered Restionales, Juncales and Cyperales as "reduced Commelinideae with reduced flowers that are mostly wind or self pollinated. All that require is that the Commelinales be ancestral to other three orders". Cronquist (1968) interpreted that the orthotropous ovules of the Commelinaceae and Restionaceae have evolved independently from the anatropous ovules of the Gramineae.

However the two families have differences both in morphological and embryological features. Morphologically Commelinaceae differ from Gramineae by possessing simple or compound helicoid cyme, usually actiomorphic flowers, biseriate perianth, trilocular ovary with one to few ovules on axile placentation and loculicidal capsules. Embryologically the family Commelinaceae differs from the Gramineae in features like Amoeboid anther tapetum, 2-celled pollen grains, orthotropous ovules, Allium and Adoxa type of embryo sacs along with Polygonum type, three ephemeral antipodals, presence of endosperm haustoria in some members like *Cyanotis axillaris*, *C. cristata* and *Commelina forskalei*.

From the above observations it is clear that the two families are not closely related to each other and reports dispel any relation between the families Commelinaceae and Gramineae.

III. Liliaceous Origin

Hallier (1912) considered that the family Gramineae has been derived from Liliaceae via Flagellariaceae. Bessey (1915) also viewed that the Gramineae is evolved from Liliales. According to Hutchinson (1934, 1973) the order Graminals along with the related orders Cyperales and Juncales is derived from Liliales. Takhtajan (1980) is also of the opinion that the order Poales have originated from the Liliales via Restionaceae whereas the order Cyperales is derived through Juncales. However the two families differ embryologically. Liliaceae differs from Gramineae in possessing usually 2-celled pollen grains at the time of anthesis, Allium, Scilla, Adoxa, Drusa, Fritillaria types of embryo sacs in addition to Polygonum type, common occurrence of hypostase, usually Helobial endosperm and Onagrad type of embryogeny along with Asterad, Caryophyllad and Chenopodiad types.

From the above embryological differences Liliaceous ancestory to Poaceae is not acceptable, though the two families in common have some similarities in their morphology.

IV. Juncales Origin

Hutchinson (1959) views of the grasses as derived from the Juncaceous stock which includes Juncaceae, Thurniaceae, Centrolepidaceae and Restionaceae. Hutchinson (1934, 1973) considered that the order Juncales is the progenitor of his Cyperales and Poales. The Juncales origin of Poaceae is also supported by Brewbaker (1967). Brewbaker (1967) considered that the order Graminales is derived from the order Juncales from which Cyperales is also originated. According to him the families Juncaceae, Restionaceae are characterised by trinucleate pollen grains whereas the family Centrolepidaceae is characterised by binucleate pollen grains. It may be assumed that the trinucleate order Graminales comprising the family Gramineae might have derived from the trinucleate families of the order Juncales while the order Cyperales consisting of both bi- and trinucleate pollen grains might have derived from the binucleate families of Juncales. The order Juncales is primarily been derived from the trinucleate Commelinales via trinucleate Liliales.

Though Hutchinson (1959; 1973) and Brewbaker (1967) supported the origin of Gramineae from Juncaceous stock embryologically these two families differ from each other. Juncaceae is characterised by the presence of simultaneous cytokinesis, usually uninucleate glandular tapetal cells whereas the Gramineae is characterised by successive cytokinesis and rarely uninucleate tapetal cells. Tetrahedral microspore tetrads are more common in Juncaceae whereas in Gramineae their occurrence is very rare and isobilateral microspore tetrads are very frequent. Ovules in Juncaceae are always anatropous while in Gramineae they may be anatropous, hemianatropous or campylotropous. The number of antipodal cells are three in Juncaceae while in Gramineae they increase in number up to 300. The type of endosperm development is of Helobial type in Juncaceae whereas it is of Nuclear type in Gramineae. The embryo development in Juncaceae is of Onagrad type while in Gramineae the development of embryo is always Asterad type.

From the above observations it is clearly evident that Juncales origin of Gramineae is not acceptable. Maze *et al.* (1970) also didn't support the hypothesis regarding the relationship of Gramineae and Juncaceae.

V. Restionales Origin

Modern ideas about the possible origin of Poaceae points to the members of the order Restionales. The workers who favoured this hypothesis include Takhtajan (1969, 1980), Cronquist (1981), Dahlgren and Clifford (1982), and Dahlgren (1983). According to Cronquist (1981) and Dahlgren (1983) the monocotyledonous families that might be considered ancestors of Poaceae are Joinvilleaceae, Restionaceae along with Flagellariaceae of the order Restionales. Maze *et al.* (1970) also observed some similarities such as proliferation of antipodals, Nuclear endosperm between Restionaceae and Gramineae. Cronquist (1981) and Dahlgren and Clifford (1982) considered Restionaceae to be the most related family to Poaceae. According to them both the families exhibit in common many sporophytic characters, additionally, a coincidence at the embryo sac organization, since they have the same type of megagametophyte in which the antipodals proliferate, the characters that has been also taken into account by Dahlgren (1982). Both the families exhibit common embryological features, such as Glandular anther tapetum, 3-celled Pollen grains, Polygonum type of embryo sac, increase in the number of antipodals and Nuclear endosperm. Based on the works of Brewbaker (1967) it is evident that the trinucleate condition of Poales is derived from the trinucleate family, Restionaceae of Juncales. Based on their work on endothecial thickenings in Poales/Restionales, Manning and Linder (1990) proposed an hypothesis which implies the evidence for a relationship between the families of Restionales/Poales. According to them the fusion of endothecial thickenings into a base plate is the basal condition which is reported both in Poales and Restionales.

Takhtajan (1969) is of the opinion that the Flagellariaceae of the Restionales is most similar to the Gramineae. Clayton (1981) points out the family Flagellariaceae as the possible origin of Poaceae. However based on the embryological features such as the presence of Allium type of embryo sac in *Flagellaria indea* and the three ephemeral antipodals which do not undergo further divisions Subramanyam and Narayana (1972) opposed such an alliance.

The family Joinvilleaceae of Restionales is little known embryologically so as to allow any consideration on their possible embryological affinities.

Based on the above observation it may be considered that the family Restionaceae of the order Restionales is one of the closest ancestral stock to Poaceae. Through at the present situation Restionaceae is considered as the closest ancestral stock to Poaceae, it is difficult to consider any strict origin to Poaceae unless otherwise the detailed embryological studies are carried out on the other related families of Poaceae.

BIBLIOGRAPHY

Adams, J.D. 1953. Observing pollen tubes within the style of *Zea mays* L. *Stain Technol.* **28**: 295–298.

Afanseva, A.S. 1956. Some new data regarding fertilization in wheat (preliminary communication). *Zh. Obshch. Biol.* **17**: 32–39.

Afanaseva, N.G. 1962. A morphological and embryological study of corn embryogenesis. *Nauk. Dokl. Vysshei Shkoly Biol. Nauk.* **4**: 107–112.

Afanaseva, N.G. 1964. An experiment in a comparative and embryological study of *Zea mays* hybrids "Tatarskaya 6" and its parental forms. *Nauk. Dokl. Vyssh. Shkoly Biol. Nauk.* **6**: 126–130.

Akerberg, E. 1939. Apomictic and sexual seed formation in *Poa pratensis. Hereditas* **25**: 359–370.

Akerberg, E. 1942. Cytogenetic studies in *Poa pratensis* and in hybrid with *Poa alpina. Hereditas* **28**: 1–126.

Akerberg, E. 1943. Further studies of the embryo and endosperm development in *Poa pratensis. Hereditas* **29**: 199–201.

Akerberg, E. and S. Bingefors. 1953. Progeny studies in the hybrid *Poa pratensis* x *Poa alpina. Hereditas* **39**: 125–136.

Alam, S. and P.C. Sandal. 1967. Cytohistological investigations of pollen abortion in male sterile sudan grass. *Crop. Sci.* **7**: 587–589.

Aleksandrov, V.G. and O.G. Aleksandrova. 1937. On the mosaic of wheat endosperm. *Dokl. Akad. Nauk. URSS* **17:**

Aleksandrova, O.G. 1937. The anatomy of various types of wheat grain. *Dokl. Akad. Nauk. URSS.* **17**: 385–388.

Alicja, H. and M. Marja. 1934. The development and degeneration of the antipodal apparatus in *Triticum durum* and *T. vulgare. Acta Soc. Bot. Polon.* **11**: 409–421.

Almgard, G. 1966. Experiments with *Poa.* III. *Ann. Agric. Coll. Sweden* **32**: 3–64.

Ambegaokar, K.B. and B.M. Johri. 1977. Seed development in Triticale–1. *Phytomorphology* **27**: 190–198.

Andersen, A.M. 1927. Development of the female gametophyte and caryopsis of *Poa pratensis* and *Poa compressa. J. Agric. Res.* **34**: 1001–1018.

Anton, A.M. 1982. Estudios sobre la biologia reproductivade *Axonopus fissifolius* (Poaceae). *Bot. Soc. Argentina* **21**: 81–130.

Anton, A.M. and A.E. Cocucci. 1984. The grass megagametophyte and its possible phylogenetic implications. *Plant syst. Evol.* **146**: 117–121.

Anton, A.M. and H.E. Connor. 1995. Floral biology and reproduction in *Poa* (Poeae: Gramineae). *Austr. J. Bot.* **43**: 577–599.

Arber, A. 1934. *The Gramineae*. Cambridge University Press, Cambridge.

Armstrong, J.M. 1937. A cytological study of the genus *Poa. Cand. J. Res. C.* **15:** 281–295.

Arthur, L., P.Ozias–Akins and W.W. Hanna. 1993. Female sterile mutant in pearl millet: evidence for initiation of apospory. *J. Heredity* **84:** 112–115.

Artschwager, E., E.W. Brandes and R.C. Starrett. 1929. Development of flower and seed of some varieties of Sugar cane. *J. Agric. Res.* **39**: 1–30.

Artschwager and C. McGuire. 1949. Cytology and reproduction of *Sorghum vulgare. J. Agric. Res.* **78**: 659–673.

Asker, S. 1979. Progress in apomixis research. *Hereditas* **91**: 231–240.

Asker, S. 1994. Genetics of apomixis mechanisms. In: *Apomixis: Exploiting hybrid vigor in rice.* G.S. Khush (ed). IRRI, Manila. p. 39–42

Asker, S., A. Hagberg and G. Hagberg. 1983. Apomixis in Barley. *Sver Utsadesforen* Tidskr. **93**: 75–76.

Asker, S.E. and L. Jerling. 1992. *Apomixis in Plants*. C.R.C. Press, Boca Raton, USA.

Assienan, B., M. Noirot and Y. Gnagne. 1993. Inheritance and genetic diversity of some enzymes in the sexual and diploid pool of the agamic complex of Maximae (*Panicum maximum* Jacq., *P. infestum* Anders. and *P. trichocladum* K. Schum.). *Euphytica* **68**: 231–239.

Aulbach–Smith, C.A. and J.M. Herr. Jr. 1984. Development of the ovule and female gametophyte in *Eustachys petrea* and *E. glauca* (Poaceae). *Amer. J. Bot.* **76**: 427–438.

Avery, G.S. Jr. 1930. Comparative anatomy and morphology of embryos and seedlings of maize, oats and wheat. *Bot. Gaz.* **89**: 1–39.

Ayyangar, G.N.R. and N. Krishnaswami. 1930. Polyembryony in *Eleusine coracana* Gaertn. Ragi. *Madras Agr. Journ.* **18**: 593–595.

Aziz, P. 1972. Histogenesis of the carpel in *Triticum aestivum. Bot. Gaz.* **133**: 376–386.

Balser, H.W. 1940. *The morphology of the flowers and inflorescences of the Cyperaceae.* Ph.D. Thesis. Cornell Univ.

Balser, H.W. 1941a Studies in the morphology of Cyperaceae. I. Morphology of flowers. A. Scirpoid genera. *Amer. J. Bot..* **28**: 542–551.

Balser, H.W. 1941b. Studies in the morphology of Cyperaceae. I. Morphology of flowers. B. Rhynchosporoid genera. *Amer. J. Bot.* **28**: 832–838.

Balser, H.W. 1944. Studies in the morphology of the Cyperaceae II. The prophyll. *Amer. J. Bot.* **31**: 53–64.

Banerjee, E.A.R. 1935. A polyembryonic grain of paddy (*Oryza sativa*). *Sci. Cult.* **1**:

Banerjee, U.C. 1967. Ultrastructure of tapetal membrane of grasses. *Grana Palynol.* **7**: 365–377.

Banerjee, U.C. and E.S. Barghoorn. 1971. The tapetal membranes in grasses and ubisch body control of mature exine pattern. In: Heslop–Harrison (ed.). Pollen development and Physiology. Butterworths, London, pp. 126–127.

Barkworth, M.E. 1982. Embryological characters and taxonomy of Stipeae (Gramineae). *Taxon* **31**: 233–243.

Basappa, G.P. and M.Muniyamma. 1981. Reproduction in two species of *Arundinella* Raddi (Poaceae). *Proc. Indian Acad. Sci. (Plant Sci.)* **90:** 477–483.

Basavaiah, 1987. *Cytomorphological studies in the genus Urochloa P. Beauv. (Poaceae)*. Ph.D. thesis, University of Mysore.

Basavaiah and T.C.S. Murthy. 1989. A cytoembryological study of *Urochloa panicoides* P. Beauv (Poaceae). *Proc. Indian Acad. Sci.* (*Plant Sci.*) **99**: 453–458.

Bashaw, E.C. 1953. *An investigation of the morphology, cytology and mode of reproduction in Cenchrus setigerus Vahl.* Ph.D. dissertation. Texas A & M University College Station.

Bashaw, E.C. 1962. Apoxixis and sexuality in Buffel grass *Pennisetum ciliare. Crop. Sci.* **2**: 412–415.

Bashaw, E.C. 1980. Apomixis and its application in crop improvement. In: *Hybridization of Crop Plants* (eds.) Fehr, W.R. and H.H. Hadley. Am. Soc. Agron and Crop. Sci. Soc., Madison, USA, p 45–63.

Bashaw, E.C. and I. Forbes, Jr. 1958a. Chromosome numbers and microsporogenesis in dallis grass *Paspalum dilatatum* Poir. *Agron. J.* **50**: 441–445.

Bashaw, E.C. and I. Forbes Jr. 1958b. Megasporogenesis, embryo sac development and embryogenesis in dallis grass, *Paspalum dilatatum* Poir. *Agron. J.* **50**: 753–756.

Bashaw, E.C. and W.W. Hanna. 1990. Apomictic reproduction. In: *Reproductive versatality in the Grasses.* G. Chapman (ed.), Cambridge Univ. Press. pp. 100–130.

Bashaw, E.C. and K.W. Hignight. 1990. Gene transfer in apomictic buffel grass through fertilization of an unreduced egg. *Crop Sci.* **30:** 571–575.

Bashaw, E.C. and B.J. Hoff. 1962. The effects of irradiation on apomictic dallis grass. *Crop Sci.* **2**: 501–504.

Bashaw, E.C. and E.C. Holt. 1958. Megasporogenesis, embryo sac development and embryogenesis in Dallis grass, *Paspalum dilatatum. Agron. J.* (USA), **50:** 753–756.

Bashaw, E.C., A.W. Hovin and E.C. Holt. 1970. Apomixis, its evolutionary significance and utilization in plant breeding. *Proc. 11th Int. Grassland Congress*, 245–248.

Bashaw, E.C., M.A. Hussey and K.W. Hignight. 1992. Hybridizaiton (n + n and 2n + n) of facultative apomictic species in the *Pennisetum* agamic complex. *Intern. J. Plant Sci.* 153: 466–470.

Battaglia, E. 1963. Apomixis. In: P. Maheshwari (ed.) *Recent Advances in the Embryology of Angiospersms.* International Society of Plant Morphologists. Univ. Delhi, Delhi. pp. 221–264.

Battaglia, E. 1992. Embryological questions: 16 unreduced embryo sacs and related problems in Angiosperms (Apomixis, cyclosis, cellularization). *Atti. Soc. Tosc. Sci. Nat. Mem. Serie.* **B 98:** 1–134.

Batygina, T.B. 1962a. Microsporogenesis and pollen grain development in – Wheat, *Triticum vulgare* var. *diamant. Dokl. Akad. Nauk SSSR Biol.* **142:** 187–190.

Batygina, T.B. 1962b. Fertilizaiton process in wheat. *Trud. Bot. Inst. Akad. Nauk. SSSR,* **5:** 260–293.

Batygina, T.B. 1966. The process of fertilization in cases of remote hybridization in the genus *Triticum. Bot. Zhur.* **51**: 1461–1470.

Batygina, T.B. 1968. Embryogensis in the genus *Triticum* (as related to the problems of Monocotyledony and remote hybridization in Gramineae). *Bot. Zhur.* **53:** 480–490 (in Russian).

Batygina, T.B. 1969. On the possibility of a new type of embryogenesis in Angiospermae. *Revue Cytol. Biol. Veg.* **32**: 335–341.

Batygina, T.B. 1971. Some peculiar features of macrosporogenesis and development of the female gametophyte in wheat. *Ann. Univ. Reims* **9:** 46–50.

Batygina, T.B. 1974a. Fertilization process in cereals. In: Linskens, H.F. (ed.) *Fertilization in Higher Plants.* Amsterdam. The Netherlands. 209–219.

Batygina, T.B. 1974b. *Wheat Embryology* (in Russian). Leningrad.

Batygina, T.B. 1978a. Embryological process at distant hybridization in Gramineae family. *Proc. Indian Natl. Sci. Acad. Part B. Biol. Sci.* **44**: 30–42.

Batygina, T.B. 1978b. Embryology of wheat. *Proc. Indian Natl. Sci. Acad. Part B. Biol. Sci.* **44**: 13–29.

Batygina, T.B., E.S. Teryokhin, G.K. Alimova and M.S. Yakovlev. 1963. Genesis of male sporangia in families Gramineae and Ericaceae. *Bot. Zh.* **48**: 1108–1120 (in Russain).

Batygina, T.B. and M.S. Yakovlev. 1990. *Comparative Embryology of Flowering Plants.* Vol. 5, Monocotyledons. St. Petersberg, Russia.

Beaudry, J.R. 1951. Seed development following the mating *Elymus virginalis* L. X. *Agropyron repens* (L.) Beauv. *Genetics* **36**: 109–133.

Bechtel, D.B. and Y. Pomeranz. 1978a. Implications of the rice kernel structure in storage, marketing and processing: A review. *J. Food Sci.* **43**: 1538–1542.

Bechtel, D.B. and Y. Pomeraz. 1978b. Ultrastructure of the mature ungerminated rice (*Oryza, sativa*) caryopsis: The germ. *Amer. J. Bot.* **65**: 75–85.

Beck, P. and J.S. Horton. 1932. Microsporogenesis and embryology in certain species of *Bromus*. *Bot. Gaz.* **93**: 42–54.

Bclk, E. 1939. *Studies in the anatomy and morphology of the spikelet and flower of the Gramineae.* Ph.D. Thesis. Cornell Univ.

Bennett, H.W. 1944. Embryology of *Paspalum dilatatum. Bot. Gaz.* **106**: 40–45.

Bennett, M.D., M.K. Rao, J.B. Smith and M.W. Bayliss. 1973. Cell development in the anther, the ovule and the young seed of *Triticum aestivum* L. var. Chinese spring. *Phil. Trans. Roc. Soc. Lond. Ser.* B **266**: 39–81.

Bensen, L. 1957. *Plant classification.* London.

Bentham, G. 1883. Cyperaceae. In: Bentham, G. and Hooker, J.D. (ed.) *Genera plantarum*. London. **3**: 1037–1073.

Bentham, G. and J.D. Hooker. 1862–1883. *Genera Plantarum*. London.

Bessey, C.E. 1915. The phylogenetic taxonomy of flowering plants. *Ann. Missouri Bot. Gard.* **2**: 109–164.

Bews, J.W. 1929. *The World's grasses. Their differentiation and distribution economics and ecology*. London.

Bhandari, N.N. and R. Khosla. 1982. Development and histochemistry of anther in Triticale cv. Tri–1. 1. Some new aspects in early ontogeny. *Phytomorphology* **32**: 18–27.

Bhanwra, R.K. 1986a. Embryological studies in five species of *Eragrostis* (Gramineae).*Res. Bull. Panjab Univ. Sci.* **37:** 17–24.

Bhanwra, R.K. 1986b. Abortive embryo sacs in *Desmostachya bipinnata. Curr. Sci.* **55**: 1033–1034.

Bhanwra, R.K. 1988. Embryology in relation to systematics of Gramineae. *Ann. Bot.* **62**: 215–233.

Bhanwra, R.K. and S.P. Choda. 1980. Comparative embryology of some grasses belonging to the tribes Aveneae, Brachypodieae, Bromeae, Festuceae and Phalaridieae. *All India Symp. on Curr. in Plant Sci.* Panjab Univ. Chandigarh *Abst.* pp. 39–40.

Bhanwra, R.K. and S.P. Choda. 1981a. Apomixis in *Eremopogon foveolatus* (Gramineae). *Nord. J. Bot.* **1:** 97–101.

Bhanwra, R.K. and S.P. Choda. 1981b. Comparative embryology of some grasses belonging to the tribes Aveneae, Brachypodieae, Bromeae, Festuceae and Phalarideae. In: *Contemp. Trends in Plant Sci.* Kalyani Publishers, New Delhi, pp 254–262.

Bhanwra, R.K. and S.P. Choda. 1984. Apomixis in Gramineae. *Res. Bull. (Sci.) Pubjab Univ.* **35**: 127–134.

Bhanwra, R.K. and S.P. Choda. 1986. Comparative embryology of *Echinochloa colonum* and *E. crusgalli* (Poaceae). *Proc. Indian Acad. Sci. (Plant Sci.)* **96**: 71–78.

Bhanwra, R.K., S.P. Choda and R. Deori. 1981. Embryological studies in some grasses. *Proc. Indian Natl. Sci. Acad.* **B. 47**: 408–418.

Bhanwra, R.K., S.P. Choda and S. Kumar. 1982. Comparative embryology of some grasses. *Proc. Indian Natl. Sci. Acad.* **B. 48:** 152–162.

Bhanwra, R.K., S.P. Choda and R. Kumar. 1985. Comparative embryology of *Brachiaria ramosa* and *Poa annua. Res. Bull. (Sci.) Punjab Univ.* **36**: 245–253.

Bhanwra, R.K., J. Kaur and A. Garg. 1991. Embryological studies in some grasses and their taxonomic significance. *Bot. J. Linn. Soc.* **107**: 405–419.

Bhanwra, R.K., S Kumar and S.P. Choda. 1980. Microsporogenesis and male gametophyte in three species of *Setaria* (Poaceae). *Curr. Sci.* **46**: 524.

Bhanwra, R.K. and P. Pathak. 1987. Embryology of *Apluda mutica. Proc. Indian Acad. Sci. (Plant Sci.)* **97**: 461–467.

Bhanwra, R.K. and D.V. Soni. 1986. The mode of reproduction in *Oplismenus burmanni, O. compositus* and *Paspalum distichum. Proc. Indian Natl. Sci. Acad.* **52:**

Bhaskara Rao, S.V. 1980. *Embryologoical studies in some forage grasses.* Ph.D. thesis, Andhra Univ., Visakhapatnam.

Bhat, P.N. and C.K. Shah. 1976. Organogenesis in the embryo of Maize, Wheat and Job's tears. In: Indo–Soviet Symposium on 'Embryology of Crop Plants'. Abst. p. 4–5.

Bhatnagar, S.P. and S. Chandra. 1975. Reproductive biology of *Triticum* III. Unfertilized ovule and embryo sac; fertilization, post fertilization changes in the embryo sac and transformation of pistil into caryopsis in relation to time. *Phytomorphology* **25**: 471–477.

Bhatnagar, S.P. and S. Chandra. 1976. Reproductive biology of *Triticum* 5. Post–pollination development of nucellus, integument and pericarp in relation to time. *Phytomorophology* **26**: 139–143.

Bhatnagar, S.O. and B.M. Johri. 1972. Development of Angiosperm seeds. In: T.T. Kozlowski. (ed.) *Seed Biology.* Vol. I. Acad. Press, New York, 47–67.

Bhojwani, S.S. and S.P. Bhatnagar. 1974. *The Embryology of Angiosperms.* Vani Educational Books. New Delhi.

Bianchi, A. 1959. Mitosi nel tappeto di antere di Mais. *Caryologia* **12**: 338–340.

Birari, S.P. 1980. Apomixis in sexuality in *Themeda* Forssk. at different ploidy levels (Gramineae). *Genetica* **54:** 133–139.

Birari, S.P. 1981. Mechanism of apomixis in *Pennisetum polystachyon* Schultz. *J. Maharashtra Agric. Univ. Pune.* **6**: 206–212.

Bjornstad, I.N. 1970. Comparative embryology of Asparagoideae Polygonateae, Liliaceae. *Nytt. Megasin fur Botanikk.* **17**: 169–207.

Bonilla, J.R. 1997. Diplosporous and aposporous apomixis in a pentaploid race of *Paspalum minus*. *Plant Science* **127**: 97–104

Borgaonkar, D.S. 1964. Apomixis and speciation in the Bothriochloininae. *Diss. Abs.* **XXV**. 2.

Bothmer, R. Von, M. Bengtsson, J. Flink and I. Linde–Laursen. 1988. Complex interspecific hybridization in Barley (*Hordeum vulgare*) and the possible occurrence of apomixis. *Theor. Appl. Gen.* **76**: 690–691.

Boyes, J.W. and W.P. Thompson. 1937. The development of the endosperm and embryo in reciprocal interspecific crosses in cereals. *J. Genet.* **34**: 203–217.

Bray, R.A. 1978. Evidence for facultative apomixis in *Cenchrus ciliaris. Euphytica* **27:** 801–804.

Bremer, G. 1946. De cytologie van soortasbastaarden bij *Saccharum. Vakblad Biologen* **21**: 3–10.

Bremer, G. 1959. Increase of chromospome number in species hybrids of *Saccharum* in relation to the embryo sac development. *Bibl. Genet.* **18**: 1–99.

Bremer, G. 1963. Problems in breeding and cytology of sugar cane. VI. Additional contemplations on fertilization and parthenogenesis in *Saccharum. Euphytica* **12**: 178–188.

Brewbaker, J.L. 1967. The distribution and phylogenetic significance of binucleate and trinucleate pollen grains in the angiosperms. *Amer. J. Bot.* **84**: 1069–1083.

Brink, R.A. and D.C. Cooper. 1944. The antipodals in relation to abnormal endosperm behaviour in *Hordeum jubatum x Secale cereale* hybrid seed. *Genetics* **29**: 391–406.

Brink, R.A. and D.C. Cooper. 1947. The endosperm in seed development. *Bot. Rev.* **13**: 423–541.

Brix, K. 1974. Sexual reproduction in *Eragrostis curvula. Z. Pflanzenzuecht.* **71:** 25–32.

Brooks, M.W., J.S. Brooks and L. Chien. 1966. The anther tapetum in cytoplasmic–genetic male–sterile *Sorghum. Amer. J. Bot.* **53**: 902–908.

Brown, R. 1814. In: Mathew Flinders, *A. Voyage to Terra Australis*. London.

Brown, W.L. 1941. The cytogenetics of *Poa pratensis. Ann. Mo. Bot. Gard.* **28**: 493–522.

Brown, W.V. 1948. Cytological study in the Gramineae. *Amer. J. Bot.* **35**: 382–395.

Brown, W.V. 1949. A cytological study of cleistogamous *Stipa leucotricha. Madrono* **10**: 97–107.

Brown, W.V. 1955. A species of grass with liquid endosperm. *Bull. Torrey Bot. Cl.* **82:** 284–285.

Brown, W.V. 1959. Apomixis in grass sub family Panicoideae. *9th Internat. Bot. Congr.* **2:** 49–50.

Brown, W.V. 1960. The Morphology of the grass embryo. *Phytomorphology* **10**: 215–223.

Brown, W.V. 1965. The grass embryo — A rebuttal. *Phytomorphology* **15**: 274–284.

Brown, W.V. and W.H.P. Emery. 1957a. Apomixis in the Gramineae, tribe Andropogoneae. *Themeda triandra* and *Bothriochloa ischaemum*. *Bot. Gaz.* **118**: 246–253.

Brown, W.V. and W.H.P. Emery. 1957b. Some South African apomictic grasses. *J. South. Afr. Bot.* **23**: 123–125.

Brown, W.V. and W.H.P. Emery. 1958. Apomixis in Gramineae. Panicoideae. *Amer. J. Bot.* **45:** 253–263.

Bruns, E. 1892. Das Grass embryo. *Flora* **76**: 1–33.

Bryant, G.D. 1952. *The apomictic mechanism* in *Bouteloua curtipendula.* Ph.D. thesis, Univ. of Oklahoma.

Burson, B.L. 1975. Cytology of some apomictic *Paspalum* species. *Crop Sci.* **16**: 229–232.

Burson, B.L. 1991. Genome relationship between tetraploid and hexaploid biotypes of dallis grass *Paspalum dilatatum. Bot. Gaz.* **152**: 219–223.

Burson, B.L. 1992. Cytology and reproductive behaviour of hybrids between *Paspalum urvillei* and two hexaploid *P. dilatatum* biotypes. *Genome* **35**: 1002–1006.

Burson, B.L. 1994. Cytology mechanisms of apospory and diplospory. In: *Apomixis : Exploiting hybrid vigor in rice.* G.S. Khush (ed), IRRI, Manila. p. 35–38

Burson, B.L. 1995. Genome relationship and reproductive behaviour of interspecific *Paspalum dilatatum* hybrids: Yellow–anthered x Uruguaina. *International Journ. Plant Sci.* **156**: 326–331.

Burson, B.L. 1997. Apomixis and sexuality in some *Paspalum* species. *Crop Sci.* **37**: 1347–1351.

Burson, B.L. and H.W. Bennett. 1970a. Cytology, method of reproduction and fertility of Brunswickgrass, *Paspalum nicorae* Parodi. *Crop. Sci.* **10**: 184–187.

Burson, B.L. and H.W. Bennett. 1970b. Cytology and reproduction of three *Paspalum* species. *J. Heredity* **61**: 129–132.

Burson, B.L. and H.W. Bennett. 1971a. Chromosome numbers, microsporogenesis and mode of reproduction of seven *Paspalum* species. *Crop. Sci.* **11**: 292–294.

Burson, B.L. and H.W. Bennett. 1971b. Meiotic and reproductive behaviour of some introduced *Paspalum* species. *J. Miss. Acad. Sci.* **17**: 5–8.

Burson, B.L. and H.W. Bennett. 1972a. Cytogenetics of *Paspalum urvillei x P. juergensii* and *P. urvillei x P. vaginatum* hybrids. *Crop Sci.* **12**: 105–108.

Burson, B.L. and H.W. Bennett. 1972b. Genome relations between an intraspecific *Paspalum dilatatum* hybrid and two diploid *Paspalum* species. *Can. J. Genet. Cytol.* **14**: 609–613.

Burson, B.L. and H.W. Bennett. 1976. Cytogenetics of *Paspalum conspersum* and its genomic relationship with yellow–anthered *P. dilatatum* and *P. malacophyllum. Can. J. Genet. Cytol.* **18**: 701–708.

Burson, B.L. and M.A. Hussey. 1998. Cytology of *Paspalum malacophyllum* and its relationship to *P. juergensii* and *P. dilatatum. Int. J. Plant Sci.* **159**: 153–159.

Burson, B.L., H. Lee and H.W. Bennett. 1973. Genome relations between tetraploid *Paspalum* dilatatum and four diploid *Paspalum species. Crop Sci.* **13**: 739–743.

Burson, B.L. and P.W. Voigt. 1996. Cytogenetic relationships between the *Eragrostis curvula* and *E. lehmanniana* complexes. *International Journal of Plant Sciences* **157**: 632–637.

Burson, B.L., P.W. Voigt and W. Evers. 1991. Cytology, reproductive behaviour and forage potential of hexaploid dallis grass biotypes. *Crop Sci.* **31**: 636–641.

Burson, B.L., P.W. Voigt, R.A. Sherman and C.L. Dewald. 1990. Apomixis sexuality in Eastern gamagrass. *Crop. Sci.* **30**: 86–89.

Burton, G.W. 1948. The method of reproduction in common Bahia grass *Paspalum notatum. J. Amer. Soc. Agron.* **40**: 443–452.

Burton, G.W. 1955. Breeding pensacola Bahia grass *Paspalum notatum*. I. Method of reproduction. *Agron. J.* **47**: 311–314.

Burton, G.W. 1982. Effect of environment on apomixis in bahia grass. *Crop. Sci.* **22**: 109–111.

Burton, G.W. and I. Forbes. 1960. The genetics and manipulation of obligate apomixis in common bahia grass (*Paspalum notatum* Flugge). *Proc. VIII Intl. Grassland Congr. Reading* (England) 66–71.

Burton, G.W. and W.W. Hanna. 1992. Using apomictic tetraploids to make a self incomplatible diploid pensacola bahia grass clone set seed. *J. Hered.* **83**: 305–306.

Burton, G.W. and J. E. Jackson. 1962. Radiation breeding of apomictic prostrate dalligrass *Paspalum dilatatum* var. *pauciciliatum. Crop. Sci.* **2**: 495–497.

Buttrose, M.S. 1963. Ultrastructure of the developing wheat endosperm. *Aust. J. Biol. Sci.* **16**: 305–317.

Cai, Xue, Mu Xi–Jin, Zhu Zhi–qring, Hau Zhi–ming, Huang Yan–hong and Sun Qu–Xin. 1997. Polyembryomy and multiple seedlings in the apomictic plants. *Acta Bot. Sinica*. **39**: 590–595.

Calderon, C.E. and T.R. Soderstrom. 1973. Morphological and anatomical considerations of the grass sub–family Bambusoideae based on a new genus *Maclurolyra. Smithson Contr. Bot.* **11**: 1–55.

Campbell, C.S. and E.A. Kellogg. 1987. Sister group relationships of the Poaceae. In: T.R. Sodestrom *et al.* (eds.) *Grass systematics and evolution.* **20:** 217–229. Smithsonian Institution Press, Washington DC.

Cannon, W.A. 1900. A morphological study of flower and embryo of wild oat, *Avena fatua. Proc. Calif. Acad. Sci.* **1:** 329–356.

Caponio, I. and C.L.Quarin. 1993. Cytology and reproduction of *Paspalum densum* and its genomic relationship with *P. intermedium* and *P. urvellei. J. Heredity* **84**: 220–222.

Carman, J.G. 1995. Gametophytic angiosperm apomicts and the occurrence of polyspory and polyembryony among their relatives. *Apomixis News letter* **8**: 39–53.

Carman, J.G. 1997. Asynchronous expression of duplicate genes in angiosperms may cause apomixis, bispory, tetraspory and polyembryomy. *Biol. J. Linn. Soc.* **61**: 51–94.

Carman, J.G., C.F. Crane and O. Riera–Lizarazu. 1991. Comparative histology of cell walls during meiotic and apomeiotic megasporogenesis in two hexaploid Australian *Elymus* species. *Crop Sci.* **31**: 1527–1532.

Carman, J.G. and S.L. Hatch. 1982. Aposporous apomixis in *Schizachyrium* (Poaceae, Andropogoneae). *Crop Sci.* **32**: 1252–1254.

Carnhan, H.L. and H.D. Hill. 1961. Cytology and genetics of forage grasses. *Bot. Rev.* **27**: 1–67.

Carniel, K. 1961. Das anther tapetum von *Zea mays. Oester Bot. Z.* **108:** 89–96.

Caro, J.A. 1982. Sinopsis taxonomica de les gramineas Argentinas. *Dominguezia* 4: 1–51.

Cartier, D. and A. Lenoir. 1980. Contribution to the study of the ovule development of *Phragmites australis. Rev. Gen. Bot.* **87**: 289–295.

Cass, D.D. 1973. An ultrastructural and normanski interference study on the sperms of barley. *Can. J. Bot.* **51**: 601–605.

Cass, D.D. and C. Fabi. 1988. Structure and properties of sperm cells isolated from the pollen of *Zea mays. Can. J. Bot.* **66**: 819–825.

Cass, D.D. and W.A. Jensen. 1970. Fertilization in Barley. *Amer. J. Bot.* **57**: 62–70.

Cass, D.D. and I. Karas. 1975. Development of sperm cells in barley. *Can. J. Bot.* **53**: 1051–1062.

Cass, D.D. and B.L. Robertson. 1979. Growth of Barley pollen tubes *in vivo*. Ultrastructural aspects of early tube growth in the stigmatic hair. *Can. J. Bot.* **57**: 386–396.

Cass, D.D., D.J. Peteya and B.L. Robertson. 1985. Magagametophyte development in *Hordeum vulgare* 1. Early megasporogenesis and nature of cell wall formation. *Can. J. Bot.* **63**: 2164–2171.

Cass, D.D., D.J. Peteya and B.L. Robertson. 1986. Megagametophyte development in *Hordeum vulgare*. 2. Later stages of wall development and morphological aspects of megagametophyte cell differentiation. *Can. J. Bot.* **64**: 2327–2336.

Cave, M.S. 1953. Cytology and embryology in the delimitation of genera. In: Plant genera, their nature and definition. *Chronica Botanica* **14**: 140–153.

Cave, M.S. 1959. Embryological characters of taxonomic value. *9th Internat. Bot. Congr.* **2**: 882.

Celarier, R.P. and J.R. Harlan. 1955. Studies on Old world Bluestems. *Oklahoma Agr. Exp. Sta. Tech. Bull.* **T. 58:** 1–31.

Celarier, R.P. and J.R. Harlan. 1956. Annual Report of Progress. Sect. I. Old world Bluestems. *Oklahoma Agr. Exp. Stat.* p.p. 1–43.

Celarier, R.P. and J.R. Harlan. 1957. Apomixis in *Bothriochloa, Dicanthium* and *Capillipedium. Phytomorphology.* **7:** 93–102. ,

Chanda, S. 1966. On the pollen morphology of the Centrolepidaceae, Restionaceae and Flagellariacae with special reference to taxonomy. *Grana palynol.* **6**: 355–415.

Chandra, N. 1962. Morphological studies in the Gramineae. *Agra. Univ. J. Res. (Sci.)* **11**: 91–92.

Chandra, N. 1963a. Morphological studies in Gramineae. IV. Embryology of *Eleusine indica* and *Dactyloctenium aegyptium. Proc. Indian Acad. Sci.* **B. 58**: 117–127.

Chandra, N. 1963b. Some ovule characters in the systematics of the Gramineae. *Curr. Sci.* **32**: 277–279.

Chandra, N. 1975. Embryology and fruit development in *Tragus biflorus. Acta Bot. Indica* **3**: 47–51.

Chandra, N. 1970. Embryology of some species of *Eragrostis*. Proc. Indian Sci. Congr. Abstract p. 249.

Chandra, N. 1976. Embryology of some species of *Eragrostis. Acta Bot. Indica* **4**: 36–43.

Chandra, S. 1970. *Post–pollination studies in Abelmoschus esculentus* (L.) *Moench and Triticum aestivum* Linn. Ph.D. Thesis, Univ. Delhi.

Chandra, S. 1976. Post pollination development of wheat caryopsis. *Indo–Soviet Symposium on Embryology of crop plants. v Abst.* 46–47.

Chandra, S. and S.P. Bhatnagar. 1974. Reproductive biology of *Triticum.* 2. Pollen germination, Pollen tube growth and its entry into the ovule – *Phytomorphology* **24:** 210–217.

Chao, C.Y. 1964. Megasporogenesis and megagametogenesis in *Paspalum orbiculare. New Asia College Annual* **6**: 15–25.

Chao, C.Y, 1971. A periodic acid – Schiffs substance related to the directional growth of pollen tube into embryo sac in *Paspalum* ovules. *Amer. J. Bot.* **58:** 649–654.

Chao, C.Y. 1974. Megasporogenesis and Megagametogenesis in *Paspalum commersonii* and *P. longifolium* at two polyploid levels. *Bot. Not.* **127**: 267–275.

Chao, C.Y. 1977. Further cytological studies of a periodic acid – Schiff's substance in the ovules of *Paspalum orbiculare. Amer. J. Bot.* **64**: 920–930.

Chao, C.Y. 1980. Autonomous development of embryo in *Paspalum conjugatum. Bot. Not.* **133**: 215–222.

Chao, S. 1970. A study on embryogeny of maize by total dissection of the embryo sac and embryo. *Acta Bot. Sin.* **17**: 328–329.

Chapman, G.P. and N. Busri. 1994. Apomixis in *Pennisetum*: An ultrastructural study. *International Journ. Plant Sci.* **155**: 492–497.

Chatterji, A.K. 1969. Apomixis and tetraploidy in *Pennisetum orientale. Crop Sci.* **9**: 796–799.

Chatterji, A.K. and G.K. Pillai. 1970. Apomixis in *Pennisetum pedicellatum. Sci. and Cult.* **36**: 667–669.

Chatterji, A.K. and D.H. Timothy. 1969. Microsporogenesis and embryogenesis in *Pennisetum flaccidum* Griseb. *Crop. Sci.* **9**: 219–222.

Chaubal, R. and B.J. Reger. 1992. Calcium in the synergid cells and other regions of pearl millet ovaries. *Sex Plant Reprod.* **5**: 34–46.

Chauhan, S.V.S. and S.P. Singh. 1976. Pollen abortion in male–sterile hexaploid wheat Norin having *Aegilops ovata* L. Cytoplasm. *Crop. Sci.* **6**: 532–535.

Chebotareu, A.A. 1970. Maize embryology: Ultrastructure of embryo sac before and at the moment of fertilization. 7e Congres Int. Micro. Electr. Grenoble **3**: 443–444.

Chen, Lanzhuang and T. Kozono. 1994a. Cytology and quantitative analysis of aposporous embryo sac development in guinea grass (*Panicum maximum* Jacq.). *Cytologia* **59**: 253–260.

Chen, Lanzhuang and T. Kozonob. 1994. Cytological evidence of seed–forming embryo development in polyembryomic ovules of facultatively apomictic guinea grass (*Panicum maximum* Jacq.). *Cytologia* **59**: 351–359.

Cheng, P.C., R.I. Greyson, Y.S. Chow and D.B. Walden. 1981. The cuticle of corn anther: Ultrastructure and waxy composition. Abst. XIII. Intl. Bot. Congr. Sydney. p. 46.

Cheng, P.C., R.I. Greyson and D.B. Walden. 1986. The anther cuticle of *Zea mays. Can. J. Bot.* **64**: 2088–2097.

Chikkannaiah, P.S. and M.S. Mahalingappa. 1975. Antipodal cells in some members of Gramineae. *Curr. Sci.* **44**: 22–23.

Chikkannaiah, P.S. and M.S. Mahalingappa. 1976a. Embryology of *Chloris gayana* (Gramineae). *J. Karnatak Univ. Sci.* **21**: 226–243.

Chikkannaiah, P.S. and M.S. Mahalingappa. 1976b. The female gametophyte and the activities of antipodals in *Cynodon dactylon.* In: *Physiology of sexual Reproduction in Flowering Plants* (ed.) C.P. Mallick *et al.* pp. 532–536. Kalyani Publishers, New Delhi.

Chikkannaiah, P.S. and M.S. Mahalingappa. 1976c. Embryological studies in *Eleusine, Indo–Soviet Symp. on Embryology of crop Plants.* New Delhi, Abst. p. 7.

Chistyokova, V.N. 1978. Irregular apomixis in breeding of wheat and incomplete 56–chromosome wheat. Couch–grass hybrids. *Proc. XIV. Int. Congr. Genet. Moscow*. Abstract. Part II. p. 162.

Cho, J. 1938. The anatomical observation of the embryo in the rice. *Bot. Mag. Tokyo.* **52**: 520–531.

Cho, J. 1956. Double fertilization in *Oryza sativa* L. and development of the endosperm with special reference to the aleurone layer. *Bull. Natn. Inst. Agric. Sci.* Tokyo: **6**: 61–101.

Choda, S.P. and R.K. Bhanwra. 1977. The mode of reproduction in *Bothriochloa odorata* and *Paspalum distichum. Proc. Indian Natl. Sci. Acad.* **43**: 175–183.

Choda, S.P. and R.K. Bhanwra. 1980. Apomixis in *Capillipedium huegelli* (Gramineae). *Proc. Indian Natl. Sci. Acad.* **B 46:** 572–578.

Choda, S.P., Harsh Mitter and R.K. Banwra. 1982. Embryological studies in three species of *Cymbopogon* Spreng. (Poaceae) *Proc. Indian Acad. Sci. (Plant Sci.)* **91**: 55–60.

Choda, S.P. and S.K. Sharma. 1981. Apomixis in *Pennisetum orientale* (Gramineae) In: *Contemp. Trends in Plant Sciences.* S.C. Verma, (ed.) Kalyani Publishers, New Delhi.

Christensen, J.E. and H.T. Horner, Jr. 1974. Pollen development and its spatial orientation during microsporogenesis in the grass *Sorghum bicolor. Amer. J. Bot.* **61**: 604–623.

Christensen, J.E., H.T. Horner and N.R. Lersten. 1972. Pollen wall and tapetal orbicular wall development in *Sorghum bicolor. Amer. J. Bot.* **56**: 43–58.

Christoff, M. 1942. Embryologische studien uber die Fortpflanzung einigher *Poa* Arten. *Jb. Univ. Heilig. Kliment Vochrid. in Sofia Fakult. f. Land u. Forstwirtsch.* **20**: 169–187.

Christoff, M.A. and R. Moskova. 1972. The apomixis and polyembryony in *Bothriochloa ischaemum* (=*Andropogon ischaemum*). *Genet. Sel.* **5:** 71–86.

Chu, C. and S.Y. Hu. 1981. The development and ultrastructure of wheat sperm cell. *XIII Intl. Bot. Cong.* Sydney. pp. 61.

Chung, S. 1974. *Meiosis and embryo sac development* in *Paspalum thunbergii.* M.Phil. diss. Chinese Univ. of Hong Kong.

Church, G.L. 1929a. Meiotic phenomena in certain Graminese 1. Festuceae, Aveneae, Agrostideae, Chlorideae and Phalarideae. *Bot. Gaz.* **87**: 608–629.

Church, .L. 1929b. Meiotic phenomena in certain Gramineae 2. Paniceae and Andropogoneae. *Bot. Gaz.* **88:** 63–84.

Clark, F.J. and F.C. Copeland. 1940. Chromosome aberrations in the endosperm of maize. *Amer. J. Bot.* **27:** 247–251.

Clausen, J., W.M. Hiesey and M.A. Nobs. 1959. Evolutionary processes in apomictic species of *Poa. Carnegie Inst. Wasting Year Book* **58**: 358–360.

Clayton, W.D. 1981. Evolution and distribution of grasses. *Ann. Missouri Bot. Gard.* **68**: 5–14.

Clifford, H.T. 1970. Monocotyledon classification with special reference to the origin of the grasses (Poaceae). *Bot. J. Linn. Soc.* **63:** 25–34.

Colhoun, C.W. and M.W. Steer. 1981. Microsporogenesis and the mechanism of cytoplasmic male sterility in maize. *Ann. Bot.* **48**: 417–424.

Combes, D. 1975. Polymorphisme et modes de reproduction dans la section des Maximae du genre. *Panicum* (Gramineae) en Afrique. Mem. ORSTOM **77**: 1–99.

Combes, D. and J. Pernes. 1970. Variations dans les nombres chromosominiques du *Panicum maximum* Jacq. enrelation avec lemode de reproduction. *C.R. Acad. Sci. Paris Series D.* **270**: 782–785.

Connor, H.E. 1974. Breeding systems in *Cortaderia* (Gramineae), *Evolution* **27**: 663–678.

Connor, H.E. 1979. Breeding systems in grasses: A survey. *New Zealand J. Bot.* **17**: 547–574.

Connor, H.E. and M.I. Dawson. 1993. Evolution of reproduction in *Lamprothyrsus* (Arundineae – Gramineae). *Annals of Missouri Bot. Gard.* **80**: 512.

Cooper, D.C. 1937. Macrosporogensis and embryo sac development in *Euchlaena mexicana* and *Zea mays. J. Agric. Res.* **55**: 539–551.

Cooper, D.C. 1951. Caryopsis development following matings between diploid and tetraploid strain of *Zea mays. Amer. J. Bot.* **38**: 702–708.

Cooper, D.C. and R.A. Brink. 1943. The antipodals as a factor in seed failure following interspecific mating in Gramineae *Records Gen. Soc. Amer.* **12**:

Costas–Lipmann, M. 1979. Embryogeny of *Cortaderia selloana* and *C. jubata* (Gramineae). *Bot. Gaz.* **140**: 393–397.

Crane, C.F. 1989. A proposed classification of apomictic mechanisms in flowering plants. *Apomixis Newsletter* **1**: 11–19.

Crane, C.F. and J.G. Carman. 1987. Mechanism of apomixis in *Elymus retisetus* from Australia and New Zealand. *Amer. J. Bot.* **74**: 477–496.

Crete, P. 1963. Embryo. In: Maheshwari, P. (ed.) *Recent advances in the embryology of Angiosperms.* International Society of Plant Morphologists. Univ. Delhi, Delhi. pp. 171–220.

Cronquist, A. 1968. *The evolution and classification of flowering plants.* London.

Cronquist, A. 1981. *An Integrated system of classification of flowering plants.* Columbia Univ. Press, New York.

Cronquist, A. 1988. *The evolution and classification of flowering plants.* 2nd ed. New York.

Cruz, R., J.W. Miles, W. Roca and H. Ramirez. 1989. Apomixis and sexuality in *Brachiaria.* I. Biochemical studies. *Cuban J. Agric. Sci.* **23**: 317–322.

Cruz, R., J.W. Miles, W. Roca and G. de LaCruz. 1989. Apomixis and sexuality in *Brachiaria* II. Cytoembryological studies. *Cuban J. Agric. Sci.* **23**: 323–328.

Cummins, M.P. 1929. Development of integument and germination of the seed of *Eleusine indica. Bull. Torrey Bot. Club.* **56**: 155–162.

Dahlgren, R.M.T. 1975. Presentation of an Angiosperm system to be used for demonstrating the distribution of characters. *Bot. Notiser* **128**: 119–147.

Dahlgren, R.M.T. 1981. Angiosperm classification and phylogeny: a rectifying comment. *J. Linn. Soc. Bot.* **82**: 89–92.

Dahlgren, R.M.T. 1983. General aspects of angiosperm evolution and macrosystematics. *Nord. J. Bot.* **3:** 119–149.

Dahlgren, R.M.T. and H.T. Klifford. 1982. *The monocotyledons: A comparative study.* London – New York, Academic Press.

Dahlgren, R. and N. Rasmussen. 1983. Monocotyledon evolution: Characters and phylogenetic estimation. In: Hecht, M.K., Wallace, B., and Prance, G.T. (ed.) *Evol. Biol.* **16**: 255:395.

Damon, E.G. 1961. Studies on the occurrence of multiploid sporocytes in three varieties of cytoplasmic male–sterile and the normal fertile variety, resistant wetland *Sorghum. Phyton* **17**: 193–203.

Das, A. and A.S. Islam. 1977. Embryological studies of three species of *Cenchrus* and their relationships. *Bangladesh J. Bot.* **6**: 123–132.

Datta, A.M. and A.K. Paul. 1951. Morphology of the ovary of *Oryza sativa* var. *plena* (double rice). *Science* **12**: 131–152.

Davis, G.L. 1966. *Systematic Embryology of Angiosperms.* John Wiley, New York.

d'Cruz, R. and P.S. Reddy. 1968. *Science & Culture* **34**: 255.

d'Cruz, R. and P.S. reddy. 1971. Inhertience of apomixis in *Dichanthium.* Indian J. *Genet. Plant. Breed.* **31**: 451–460.

den Nijs, A.P.M. and Van G.E. Dijk. 1993. In: *Plant Breeding principles and prospects* (eds.) Hayward, M.D., N.O. Bosemark and I. Romagosa. Chapman and Hall, London pp. 229–245.

Deori, R. 1981. Embryological studies in some grasses. *Proc. Indian Acad. Sci.* **B. 47**: 408–418.

Deshpande, P.K. 1976. Development of embryo and endosperm in *Eragrostis unioloides* (Poaceae). *Plant Syst. Evol.* **125**: 253–259.

Deshpande, P.K. 1981. Induced male sterile mutants in Barley var. clipper. *Indian J. Exp. Biol.* **19**: 277–279.

Deshpande, P.K. and K.H. Makde. 1995. Embryo and fruit in Poaceae. In: Y.S. Chauhan and A.K. Pandey (eds.). Advances in Plant Reproductive Biology. Vol. 1, pp. 101–116.

Deshpande, P.K. and P.S.G. Raju. 1979. Development of caryopsis and localization of DNA proteins in *Triticum* species. *Phytomorphology* **29**: 100–111.

Deshpande, P.K. and P.S.G. Raju. 1985. Meiotic behaviour and microspore tetrad patterns in *Triticum species.* In: C.M. Govil and V. Kumar (eds.) *Trends in Plant Research.* Dehra dun. pp. 188–193.

De Triquell, A.A. 1987. Grass gametophytes, their origin, structure and relation with the sporophyte. In: T.R. Soderstrom, K.W. Hilu, C.S. Campbell and M.E. Barkworth (eds.). *Grass Systematics and Evolution*, 11–20. Smithsonian Institution Press, Washington, D.C.

Dewald, C.L. and B. Kindiger. 1994. Genetic transfer of gynomonoecy from diploid to triploid Eastern gama grass. *Crop Sci.* **34**: 1259–1262.

Dewald, C.L. and B.K. Kindiger. 1998. Cytological and molecular evaluation of the reproductive behaviour of *Tripsacum andersonii* and a female fertile derivative (Poaceae). *Amer. J. Bot.* **85**: 1237–1242.

De Wet, J.M.J. 1965. Diploid races of tetraploid *Dichanthium* species. *Am. Nat.* **99**: 167–171.

De Wet, J.M.J. 1968. Diploid–tetraploid–haploid cycles and the origin of variability in *Dichanthium* agamospecies. *Evolution* **22**: 394–397.

De Wet, J.M.J. 1971. Reversible tetraploidy as an evolutionary mechanism. *Evolution* **25**: 545–548.

De Wet, J.M.J. 1978. Systematics and evolution of *Sorghum* Sec. *Sorghum* (Gramineae) *Amer. J. Bot.* **65**: 477.

De Wet, J.M.J. and D.S. Borgaonkar. 1963. Aneuploidy and apomixis in *Bothriochloa* and *Dicanthium*. *Bot. Gaz.* **124**: 437–440.

De Wet, J.M.J. and J.R. Harlan. 1970. Apomixis, polyploidy and speciation in *Dichanthium*. *Evolution* **24**: 270–277.

De Wet, J.M.J., D.E. Brink and C.E. Cohen. 1983. Systematics of *Tripsacum*: Section *Fasciculata* (Gramineae). *Amer. J. Bot.* **70**: 1139–1146.

De Wet, J.M.J., J.R. Harlan Jr. and D.E. Brink. 1982. Systematics of *Tripsacum dactyloids*. *Amer. J. Bot.* **69**: 1251–1257.

Diboll, A.G. 1964. Electron micronscopy of female gametophyte and early embryo development in maize. Ph.D. Thesis Univ. of Texas.

Diboll, A.G. 1968a. Histochemistry and fine structure of the pollen tube residue in the megagametophyte of *Zea mays*. *Caryologia* **21**:

Diboll, A.G. 1968b. Fine structural development of the megagametophyte of *Zea mays* following fertilization. *Amer. J. Bot.* **55**: 787–806.

Diboll, A.G. and D.A. Larson. 1966. An electron microscopic study of the mature megagametophyte in *Zea mays*. *Amer. J. Bot.* **53**: 391–402.

Di Fulvio, T. 1981. Embryology in the Systematics of Angiosperms. *Kurtziana* **14**: 21–40.

Disk, G.E. Van and G.D. Winkelhorst. 1982. Interspecific crosses as a tool, in breeding *Poa pratensis* L.I. *Poa longifolia* Trin. x *P. pratensiss* L. *Euphytica* **31**: 215–223.

Diwanji, V.B. 1973. The gametophytes of *Eragrostis poaeoides*. Beauv. Proc. 63rd *Indian Sci. Congr.* Abstract p. 328.

Diwanji, V.B. 1976. *Embryological studies in the Gramineae.* Ph.D. Thesis, Univ. Indore, M.P., India.

Diwanji, P.V. and V.B. Diwanji. 1981. Life history of *Oropetium thomaeum* and *O. villosulum*. 4th All India Bot. Conf. Abstr. p. 58.

Diwanji, V.B,. and M.D. Padhye. 1977. The life history of *Eragrostis tenuifolia.* In: *Recent Trends and Contacts between Cytogenetics Embryology and Morphology* (ed.) V.R. Dnyansagar p. 261–276.

Dore, W.G. 1956. Some grass genera with liquid endosperm. *Bull. Torrey Bot. Cl.* **83**: 335–337.

Do Valle, C.B. and C. Glienke. 1991. *Apomixis News Letter 3*: 11

Do Valle, C.B. and C.B. Miles. 1992. *Apomixis News Letter 5*: 37

Dronzek, B.L., P. Hwang and W. Bushuk. 1972. Scanning electron microscopy of starch from sprouted wheat. *Cereal Chem.* **49**: 232–239.

Dujardin, M. and W.W. Hanna. 1983a. Meiotic and reproductive behaviour of facultative apomictic BC offspring derived from *Pennisetum americanum – P. orientale* interspecific hybrids. *Crop Sci.* **23:** 156–160.

Dujardin, M. and W.W. Hanna. 1983b. Apomictic and sexual pearl millet x *Pennisetum squammulatum* hybrids. *J. Hered.* **74**: 277–279.

Dujardin, M. and W.W. Hanna. 1984. Microsporogenesis, reproductive behaviour and fertility in five *Pennisetum* species. *Theor. Appl. genet.* **67:** 197--202.

Dujardin, M. and W.W. Hanna. 1988. Production of 27, 28 and 56–chromosome apomictic hybrid derivatives between pearl millet (2n = 14) and *Pennisetum squammulatum* (2n = 54). *Euphytica* **38:** 229–235.

Dujardin, M. and W.W. Hanna. 1989a. Developing apomictic pearl millet: Characterisation of a BC_3 plant. *J. Genet. Breed.* **43**: 145–150.

Diyardin, M. and W.W. Hanna. 1989b. Fertility improvement in tetraploid pearl millet characterisation of a BC_3 plant. *J. Genet. Breed.* **43**: 145–151.

Dujardin, M. and W.W.Hanna. 1994. Transfer of alien chromosome carrying gene for apomixis to cultivated *Pennisetum*. In: *Apomixis : Exploiting hybrid vigor in rice.* G.S. Khush (ed), IRRI, Manila p. 31–34.

Dumas, C. and H.L. Mogensen. 1993. Gametes and fertilization: maize as a model system for experimental embryogenesis in flowering plants. *The Plant Cell* **5**: 1337–1348.

Dzevaltovskii, A. and S.P. Skpilevaya 1978. *Byull. Gl. Botan. Sada. An SSSR* **19**: 60.

Emery, W.H.P. 1957. A study of reproduction in *Setaria macrostachya* and its relatives in the south–western United States and Northern Mexico. *Bull. Torrey Bot. Cl.* **84**: 106–121.

Edward, T.E. 1971. Survey of occurrence of liquid or soft endosperm in grass genera. *Bull. Torrey Bot. Club* **98:** 264–268.

El–Ghazaly, G. and W.A. Jensen. 1985. Studies of the development of wheat *(Triticum aestivum)* pollen. III. Formation of microchannels in the exine. *Pollen Spores* **27**: 5–14.

El–Ghazaly, G. and W.A. Jensen. 1986. Studies of the development of wheat *(Triticum aestivum)* pollen. I. Formation of the pollen wall and ubisch bodies. *Grana* **25**: 1–29.

El–Ghazaly, G. and W.A. Jensen. 1987. Development of wheat (*Triticum aestivum*) pollen. I. Histochemical differentiation of wall and ubisch bodies during development. *Amer. J. Bot.* **74**: 1396–1418.

El–Ghazaly, G. and W.A. Jensen. 1990. Development of wheat (*Triticum aestivum*) pollen wall before and after effect of gametocide. *Canad. J. Bot.* **68**: 2509–2516.

Elgin, J.H. and J.P. Miksche (eds.). 1992. *Proceedings of the apomixis workshop.* U.S. Dept. of Agriculture, ARS–104, 64 pp.

Elkonin, L.A., N.K. Enaleeva, M.I. Tsvetova, E.V. Belyayeva and A.G. Ishin. 1995. Partially fertile line with apospory obtained from tissue culture of male sterile plant of sorghum (*Sorghum bicolor*). *Annals of Botany*, **76**: 359–364.

Emery, W.H.P. and W.V. Brown. 1957. Extra–ovular development of embryos in two grass species. *Bull. Torrey Bot. Cl.* **84**: 3761–365.

Emery, W.H.P. and W.V. Brown. 1958. Apomixis in Gramineae, tribe Andropogoneae, *Heteropogon contortus. Madrono* **14**: 238–246.

Engelbert, V. 1940. Reproduction in some *Poa* species. *Canad. J. Res.* **18**: 518–521.

Engelbert, V. 1941. The development of twin embryo sacs, embryos and endosperm in *Poa arctica. Canad. J. Res.* **19:** 135–144.

Emery, W.H.P. and M.N. Guy. 1979. Reproduction and embryo development in Texas wild rice (*Zizania texana* Hitchc). *Bull. Torrey Bot.* Cl. **106:** 29–31.

Engell, K. 1989. Embryology of Barley: time course and analysis of controlled fertilization and early embryo formation based on serial sections. *Nord. J. Bot.* **9**: 265–280.

Engle, L.M., J.M.J. de Wet and J.R. Harlan. 1974. Chromosomal variation among offspring of hybrid derivatives with 20 *Zea* and 36 *Tripsacum* chromosomes. *Caryologia* **27**: 193–209.

Engler, A. 1887. *Hochgebrigspflanze.* Leipzig.

Engler, A. 1926. *Die Naturlichen pflanzen familian.* Leipzig.

Engler, A. and L. Diels. 1936. *Syllabus der Pflanzenfamilien.* Leipzig.

Engler, A. and K. Prantl. 1887–1915. *Die Naturlichen Pflanzenfamilien.* **20** vols. Wilhelm Engelmann, Leipzig.

Erdelska, O. 1966. Influence of fertilization on the development of antipodal nuclei in Barley. *Biologia (Bratislava)* **21:** 857–864.

Erdelska, O. 1967. Polar nuclei in unfertilized barley embryo sacs. *Ann. Bot.* **31**: 367–369.

Evans, L.T. and R.B. Knox. 1969. Environmental control of reproduction in *Themeda anstralis. Austral. J. Bot.* **17**: 375–389.

Evers, A.D. 1970. Development of endosperm of wheat. *Amer. J. Bot.* **34**: 547–555.

Farquharson, L.I. 1954. Apomixis, polyembryony and related problems in *Tripsacum.* Thesis. Univ. Microfilms. Ann. Arbor. Mich.

Farquharson, L.I. 1955. Apomixis and polyembryony in *Tripsacum dactyloides. Amer. J. Bot.* **42**: 737–743.

Faruqi, S.A., S. Afridi and S.M.H. Usmani. 1975. Apomixis and polyembryony in some Andropogoneae from Pakistan. *Pak. J. Bot.* **7:** 135–138.

Faure, J.E., C. Diagonnet and C. Dumas. 1994. An *in vitro* system for adhesion and fusion of maize gametes. *Science* **263**: 1598–1600.

Faure, J.E., H.L. Mogensen, C. Dumas, H. Lorz and E. Kraz. 1993. Karyogamy after electrofusion of single egg and sperm cell protoplasts from maize: cytological evidence and time course. *The plant cell* **5:** 747–755.

Febulaus, G.N.V. 1992. Apomixis in *Panicum notatum* Retz. (Poaceae). *J. Indian Bot. Soc.* **71**: 147–151.

Febulaus, G.N.V. and T. Pullaiah. 1990. Embryology of *Eragrostis viscosa*. *Ann. Bot.* (Roma), **48**: 49–56.

Febulaus, G.N.V. and T. Pullaiah. 1991a. Embryology of *Chloris roxburghiana*. *Taiwania* **36**: 303–310.

Febulaus, G.N.V. and T. Pullaiah. 1991b. Embryology of *Brahiaria reptans.* In: N.C. Aery and B.L. Chaudhary (ed.). *Botanical Researches in India.* Udaipur p. 149–143.

Febulaus, G.N.V. and T. Pullaiah. 1992a. Embryological studies in *Panicum repens* Linn. *Phytomorphology* **42**: 125–131.

Febulaus, G.N.V. and T. Pullaiah. 1992b. Embryology of *Eragrostiella bifaria* (Vahl) Bor (Poaceae). *Ann. Bot.* (Roma) **50**: 5–15.

Febulaus, G.N.V. and T. Pullaiah. 1993a. Embryological studies in *Perotis* (Poaceae). *Acta Bot. Ind.* **21**: 78–81.

Febulaus, G.N.V. and T. Pullaiah. 1993b. Embryology of *Aristida* (Poaceae). *Taiwania* **38**: 38–48.

Febulaus, G.N.V. and T. Pullaiah. 1995. Apomixis in *Cenchrus ciliaris*. In: *Advances in Plant Reproductive Biology* (eds.) Y.S. Chauhan and A.K. Pandey. Narendra Publishing House, Delhi. pp. 153–161.

Febulaus, G.N.V. and T. Pullaiah. 1996. Embryology of *Digitaria* (Poaceae). *Phytomorphology*. **46**: 161–169.

Febulaus, G.N.V. and T. Pullaiah. 1997. Embryology of *Melanocenchris jacquemontii* Jaub. ex Spach. *J. Plant Anat. Morph.* **7:** 159–165.

Fineran, B.A., D.J.C. Wild and M. Ingerfeld. 1982. Initial wall formation in the endosperm of wheat *Triticum aestivum*: A re–evaulation. *Can. J. Bot.* **60**: 1776–1795.

Fischer, A. 1880. Zur Kenntnis der embryo sac entwicklung einiger Angiospermen. *Jen. Zeit. f. Nat.* (7) **14**: 9–132.

Fischer, W.D. 1953. *A study of the morphology, cytology and breeding behaviour of Pennisetum ciliare*. Ph.D. dissertation. Texas A & M Univ. College Station.

Fisher, W.D., E.C. Bashaw and E.C. Holt. 1954. Evidence for apomixis in *Pennisetum ciliare* and *Cenchrus setigerus. Agron. Journ.* **46:** 401–404.

Focke, W.O. 1881. "Die Samenban bei *Cynastrum. Svensk Bot. Tidskr.* **13**: 295–304.

Freter, L.E. and V.W. Brown. 1995. A cytotaxonomic study of *Bouteloua curtipendula* and *B. uniflora. Bull. Torrey bot. Cl.* **82**: 118–130.

Funk, C.R. and S.J. Han. 1967. Recurrent interspecific hybridization: a proposed method of breeding kentucky blue grass *Poa pratensis. New Jersy Agric. Expt. Sta. Bull.* 818: 3–14.

Furness, C.A. and P.J. Rudall. 1998. The tapetum and systematics in monocotyledons. *Bot. Rev.* **64**: 201–239.

Furssov, V.I. 1961. Some cytochemical data on the sexual process and embryogenesis of wheat. *Nauchn. Dokl. Mosk. Sel'skokhoz. Kazakhask SSSR* **2**: 17–26.

Gaines, E. and H. Aase. 1926. A haploid wheat plant. *Amer. J. Bot.* **13**: 373–385.

Gavrilova, O.M. 1961. Certain features of the growth of pollen tubes in wheat as related to the phenomenon of multiple fertilization. *Trudy Inst. Genet.* **28**: 200–207.

Gavrilova, O.M. 1962. Concerning the possible realisation of additional pollen tubes and sperms in the wheat ovary. *Trudy Inst. Genet.* **29**: 231–237.

Gawali, S.P. 1990. Embryological studies in the Gramineae. II. Embryogeny in *Brachiaria eruciformis* (Sm.) Griseb and *Paspalidium punctatum* (Burm.) A. Camus. *12th All India Bot. Conf. Abstract.*

German, J.G. and S.L. Hatch. 1982. Aposporous apomixis in *Schizachyrium* (Poaceae, Andropogoneae). *Crop Sci.* **22**: 1252–1254.

Ghaisas, V.A. and B.M. Patil. 1990. Embryology of *Eragrostic namaquensis* Schrad. var. *diplochnoides* (Steud.) W.D. Clayton with a brief discussion on organisation of embryo. 12th All India Bot. Conf. abstract.

Gildenhuys, P.J. and K. Brix. 1958. Cytological abnormalities in *Pennisetum dubium. Heredity* **12**: 441–452.

Gildenhuys, P.S. and K. Brix. 1959. Apomixis in *Pennisetum dubium. S. Afr. J. Agr. Sci.* **2:** 231–245.

Gildenhuys, P.S. and K. Brix. 1965. The relationship between embryo and endospem in interspecific hybrids in *Pennisetum, Ann. Bot.* **29**: 709–715.

Gioelli, F. 1935. Ricerche embryologische sulla *Phyllostachya nigra. Annalli di Botanica* **21**: 51–60.

Gledhill, D. 1967. Embryo sac formation in African *Axonopus* species. *Phytomorphology* **17**: 214–223.

Gobbe, J., B. Longly and B.P. Louant. 1982. Calendrier des sporogeneses et gametogeneses femelles chez lo diploide et le tetraploide induit de *Brachiaria ruziziensis. Can. J. Bot.* **60:** 2031–2036.

Godbole, R.K. 1968. *Embryological studies in some cultivated varieties of Sorghum.* M.Sc. Diss. Coll. of Agricult. Nagpur.

Gordon, N. 1922. The development of endosperm in cereals. *Proc. Roy. Soc. Victoria.* **34**: 105–106.

Goring, H. 1960. Untersuchungen zum Befruchungsprozess bei *Zea mays* mit radioaktiven Phosphor. *Naturwissenschaften.* **47**: 142.

Gould, F.W. 1968. *Grass Systematics.* McGraw Hill. New York.

Gould, F.W. 1970. Linear microspore tetrads in the grass *Stipa ischu. Madrono.* **20**: 411–412.

Gounaris, E.K., R.T. Sherwood, I. Gounaris, R.H. Hamilton and D.L. Gustine. 1991. Inorganic salts modify embryo sac development in sexual and aposporous *Cenchrus ciliaris. Sex Plant Reprod.* **4**: 188–192.

Grazi, F., M. Umaerus and E. Akerberg. 1961. Observations in the mode of reproduction and the embryology of *Poa pratensis. Hereditas* **46**: 489–541.

Greene, C.W. 1984. Sexual and apomictic reproduction in *Calamogrostis* (Gramineae) from Eastern North America. *Amer. J. Bot.* **71**: 285–293.

Grimanelli, D., O. Leblanc, E. Espinosa, E. Paroth, D. Gonzalez de Leon and Y. Savidan. 1998a. Mapping diplosporous apomixis in tetraploid *Dipsacum*: One gene or several genes. *Heredity* **80**: 33–39.

Grimanelli, D., O. Leblanc, E.. Espinosa, E. Perotti, D. Gonzalez de Leon and Y. Savidan. 1998b. Non mendelian transmission of apomixis in maize–*Tripsacum* hybrids caused by a transmission ratio distortion. *Heredity* **80:** 40–47.

Groeber, K. 1986. Cytological and embryological studies in nucelli of apomictic specimens of *Poa pratensis* L. *Kulturpflanze.* **34**: 147–158.

Grober, K., F. Matzk and M. Zacharias. 1974. Untersuchungen zur entwicklung der apomiktischen Fortpflanzungs — Weise bei Futtergrasern. I. Art und Gattungskreuzungen. *Kulturflanze* **22**: 159–180.

Grober, K., F. Matzk and M. Zacharias. 1976. Untrsuchungen zur entwicklung der apomiktischen Fortpflanzungs Weise bei Futtergrassen. II. Hybrid–effekt und Fertilitat von Art–und Gattungsbastarden. *Kulturflanze* **24**: 349–364.

Grun, P. 1954. Cytogenetic studies of *Poa*–I. Chromosome numbers and morphology of interspecific hybrids. *Amer. J. Bot.* **42**: 671–678.

Grun, P. 1955. Cytogenetic studies of *Poa*–III. Variation within *Poa nervosa*, an obligate apomict. *Amer. J. Bot.* **42**: 778–784.

Grun. P. 1975. Du cotyledon des monocotyledons. *Phytomorphology,* **25**: 193–200.

Guignard, J.L. 1961a. Embryogenic des Graminees/ Development de l'embryon chez le *Setaria verticillata. Bull. Soc. Bot.* France **108**: 212–217.

Guignard, J.L. 1961b. Recherches sur l'embryogenie des Graminees, rapports des Graminees avec les autres Monocotyledones. *Ann. Sci. Nat. Ser.* **2**: 491–610.

Guignard, J.L. 1962. Recerches sur l'embryogenie des Graminees. Rapport des Graminees avec les autres Monocotyledones. *Ann. Sci. Nat. Bot. Ser.* **12**: 2: 491–610.

Guignard, J.L. 1963. Development de l'embryon chez le *Mibora minima. Bull. Soc. Bot.* France **110**: 193–194.

Guignard, J.L. and J.C. Mestre. 1970. L'embryon des Graminees. *Phytomorphology* **20:** 190–197.

Gupta, P.K. 1968. Observations on degree of aprospory in three members of Andropogoneae. *Curr. Sci.* **37**: 295–296.

Gupta, P.K. 1969–70. Apomixis in *Bothriochloa pertusa* (L.) A. Camus. *Port. Acta Biol. Ser.* **A. 11**: 279–287.

Gupta, P.K., R.P. Roy and A.P. Singh. 1969. *Portung Acta Biol.* **1:** 253.

Gupta, P.K. and Yashvir. 1971. Apomixis in *Cenhrus ciliaris* — A preliminary study. *Curr. Sci.* **40**: 444–446.

Gustaffson, A. 1946. Apomixis in higher plants. I. The mechanism of Apomixis. *Lunds Univ. Arsskr.* **42**: 1–66.

Gustine, D.L., R.T. Sherwood, Y. Gounaris and D. Huff. 1996. Isozyme, protein and RAPD markers within a half sibfamily of buffelgrass segregating for apospory. *Crop. Sci.* **36**: 723–737.

Gymer, P.T. and W.V. Whittington. 1975a. Hybrids between *Lolium perenne* and *Festuca pratensis*. Cytological abnormalities. *New Phytol.* **74**: 295–306.

Gymer, P.T. and W.V. Whittington. 1975b. Hybrids between *Lolium perenne* and *Festuca pratensis*. Meiosis and fertility. *New Phytol.* **75**: 259–267.

Hackansson, A. and S. Ellerstrom. 1950. Seed development after reciprocal crosses between diploid and tetraploid rye. *Hereditas* **36**: 256–296.

Haeckel, E. 1887. Echte Graser. Engler and Prantl. *Des Naturlichen Pflanzenfamilien* II. 2.

Hagberg. G. and A. Hagberg. 1981. Haploidy initiator gene in barley. *Proc. X Intl. Barley Gen. Sympos. Edinburgh,* pp. 686–689.

Hair, J.B. 1956. Subsexual reproduction in *Agropyron. Heredity.* **19**: 129–160.

Hakansson, A. 1943. Die Entwicklung des Embryo sacks and die Befruchtung bei *Poa alpina. Hereditas* **29**: 25–61.

Hakansson, A. 1944. Erganende Beitrage zur Embryologie von *Poa alpina. Bot. Notiser* **1944**: 299–311.

Hakansson, A. 1946. Some observations on the seed development in two strains of *Triticale. Acta Agric. Suecl.* **1:** 377–384.

Hakansson, A. 1948. Embryology of *Poa alpina*. Plants with accessory chromosomes. *Hereditas* **34**: 233–247.

Hallier, H. 1912. L'origine et le system phyletique der Angiospermes. *Arch. Neerl. Sci. Exact. Nat. Ser. III.* **B: 1**: 146–234.

Hanna, W.W. and E.C. Bashaw. 1987. Apomixis: Its identification and use in plant breeding *Crop. Sci.* **27**: 1136–1139.

Hanna, J.R., M.H. Brooks, D.S. Borgaonkar and J.M.J. de Wet. 1964. Nature and inheritance of apomixis in *Bothrichloa.*

Hanna, W.W. 1970. Apospory in *Sorghum bicolor* (L.) Molnch. *Science* **1970**: 338–339.

Hanna, W. 1995. Use of apomixis in cultivar development *Adv. Agr.* **54**: 343–350.

Hanna, W.W., M.Dujardin, P. Ozias–Akins and L. Arthur. 1993. Transfer of apomixis in *Pennisetum.* In: Intern. Workshop on Apomixis in rice. K.J. Wilson (ed.). Rockfeller — Foundation and China Ntl. Ctr. Biotech. Dev. pp. 23–27.

Hanna, W.W., M. Dujardin, P. Ozias–Akins, E.Lubbers and L.Arthur. 1993. Reproduction, cytology and fertility of pearl millet x *Pennisetum squammulatum* BC 4 plants. J. Heredity **84:** 213–216.

Hanna, W.W. and J.B. Powell. 1973. Stubby head, an induced facultative apomict in pearl millet. *Crop Sci.* **13**: 726–728.

Hanna, W.W., J.B. Powell and G.W. Burton. 1976. Relationship to polyembryony, frequency, morphology, reproductive behaviour and cytology of antotetraploids in *Pennisetum americanum. Can. J. Genet. Cytol.* **18**: 529–536.

Hanna, W.W., K.F. Schertz and E.C. Bashaw. 1970. Apospory in *Sorghum bicolor* (L.) Monench. *Science* **170:** 338–339.

Harigopal, B. 1982. *Reproductive biology of some Indian bamboos.* Ph.D. Thesis, Univ. of Delhi.

Harigopal, B. and Manasi Ram. 1981. Floral morphology, development of sporangia and sporogenesis in *Dendrocalamus hamiltonii. Proc. Indian Natl. Sci. Acad. Part B. Biol. Sci.* **17**: 519–526.

Harigopal, B. and H.Y. Mohan Ram. 1985. Systematic significance of mature embryo of bamboos. *Plant Syst. Evol.* **14B**: 239–246.

Harlan, H.V. and M.N. Pope. 1925. Some cases of apparent single fertilization in Barley. *Amer. J. Bot.* **12**: 50–53.

Harlan, J.R. 1949. Apomixis inside Oats gramma. *Amer. J. Bot.* **36**: 495–499.

Harlan, J.R., M.H. Brooks. D.S. Borgaonkar and J.M.J. de Wet. 1964. Nature and inheritance of apomixis in *Bothriochloa* and *Dicanthium. Bot. Gaz.* **125**: 41–46.

Harlan, J.R., R.P. Celaries, W.L. Richardson, M.H. Brooks and K.L. Mehra. 1958. Studies on Old World Bluestems. II. *Okla. Agric. Exp. Sta. Techn. Bull. T.* **72**: 3–23.

Harlan, J.R. and J.M.J. deWet. 1963. Role of apomixis in the evolution of the *Bothriochloa–Dichanthium* complex. *Crop Sci.* **3**: 314–316.

Harlan, J.R., J.M.J. de Wet and C.A. Newell. 1978. Apomixis and pseudo–apomixis in *Tripsacum.* Abstr. Proc. XIV Intl. Congr. Genet. Moscow. Vol. II, p. 166.

Hayward, M.P. and M.A.P. Mantiratna. 1972. Pollen development and variation in the genus *Lolium* 1. Pollen size and tapetal relationships in male fertile and male sterile lines. *Z. Pflanzenzeucht.* **67**: 131–144.

Herr, J.M. Jr. 1984. Embryology and Taxonomy. In: Johri, B.M. (ed.) *Embryology of Angiosperms*: Springer–Verlag. Berlin.

Hegnauer, R. 1963. *Chemotoxonomie der Pflanzen.* Band. 2, Monocotyledoneae. Basel.

Heslop–Harrison, J. 1979. Aspects of the structure, cytochemistry and germination of the pollen of rye (*Secale cereale* L.). *Ann. Bot.* **44** (Suppl.) 1–47.

Heslop–Harrison, J. and Y. Heslop–Harrison. 1980a. The pollen–stigma interaction in the grasses. 1. Fine structure and cytochemistry of stigmas of *Hordeum* and *Secale. Acta Bot. Neerl.* **29**: 261–276.

Heslop–Harrison, J. and Y. Heslop–Harrison. 1980b. Cytochemistry and function of the zweischenkorper in grass pollens. *Pollen Spore* **22**: 4–10.

Heyman, D.L. 1956. Apomixis in Australian *Paspalum dilatatum. Aust. Inst. Agri. Sci. J.* 242–293.

Hignight, K.W., E.C. Bashaw and M.A. Hussey. 1991. Cytological and morphological diversity of native apomicitic buffel grass *Pennisetum ciliare* (L.) Link. *Bot. Gaz.* **152**: 214–218.

Hilu, K.W. and K. Wright. 1982. Systematics of Gramineae. A cluster analysis study. *Taxon* **3**: 9–36.

Hiroyoshi, I. 1938. Morphology of pollen nuclei and their behaviour at germination in rice plant. *Proc. Crop. Sci. Soc. Jap.* **10**: 65–70.

Hoshikawa, K. 1959. Cytological studies of double fertilization in wheat (*Triticum aestivum* L.). *Proc. Crop. Sci. Soc. Japan.* **34**: 142–144.

Hoshikawa, K. 1960a. Observations on the embryo sac containing double egg apparatus in *Triticum aestivum* L. *Bot. Mag.* Tokyo. **73**: 107–112.

Hoshikawa, K. 1960b. Influence of temperature upon the fertilization wheat, grown in various levels of nitrogen. *Proc. Crop. Sci. Soc. Japan.* **28**: 291–295.

Hoshikawa, K. 1961a. Studies on the ripening of wheat grain. 1. Embryological observation of the early development of the endosperm. *Proc. Crop. Sci. Soc. Jap.* **29**: 253–257.

Hoshikawa, K. 1961b. Studies on the ripening of wheat grain. 2. Development of endosperm tissue. 3. Development of starch grain and reserve protein particle in the endosperm. *Proc. Crop. Sci. Soc. Jap.* **29**: 253–257.

Hoshikawa, K. and A. Higuchi. 1960–61. Embryo sac formation in wheat. *Proc. Crop. Sci. Soc. Jap.* **29**: 107–113.

Hu, S.Y. 1964. Morphological and cytological observations on the process of fertilization in wheat. *Scientia Sin.* **13**: 925–936.

Hu, Gongshe, G.H. Liang and C.E. Wassom. 1991. Chemical induction of apomictic seed formation in maize. *Euphytica* **56**: 97–106.

Huff, D.R. and J.M. Bara. 1993. Determining genetic origins of aberrant progeny from facultative apomictic kentucky blue grass using a combination of flow cytometry and silver stained RAPD markers. *Theor. Appl. Genet.* **87**: 201–208.

Hussey, M.A., E.C. Bashaw, K.W. Hignight and M.L. Dahmer. 1991. Influence of photoperiod on the frequency of sexual embryo sacs in facultative apomictic buffel grass. *Euphytica* **54**: 141–146.

Hutchinson, J. 1934. *The families of flowering plants.* II. *Monocotyledons.* London.

Hutchinson, J. 1948. *British flowering plants*, London.

Hutchinson, J. 1959. *Evolution and phylogeny of flowering plants.* London.

Hutchinson, J. 1969. *Evolution and phylogeny of flowering plants.* London.

Hutchinson, D.J. and E.C. Bashaw. 1964. Cytology and reproduction in *Panicum coloratum* and related species. *Crop Sci.* **4**: 151–153.

Huysmans, S., G. El–Ghazaly and E. Smets. 1998. Orbicules in Angiosperms: Morphology, function, distribution and relation with tapetum types. *Bot. Rev.* **64**: 240–272.

Indira Devi, P. 1967. *Embryological studies in the families Gramineae and Cyperaceae.* Ph.D. Thesis, Andhra Univ., Visakhapatnam.

Islam, A.S. and A. Das. 1970. Apomixis in *Cenchrus pennisetiformis* Hochst. *Proc. 21–22 Pak. Sci. Conf.* Rajashahi Part III.

Ivanovskaja, H.V. 1973. Functional morphology of polytene chromosomes of wheat antipodal cells. *Tsitologia* **15**: 1445–1452.

Jacobson, J.V., R.B. Knox and N.A. Pyliotis. 1971. The structure and composition of aleurone grains in the barley aleurone layer. *Planta* **101**: 189–209.

Jane, wan–Neng. 1997. Ultra–structure of the maturing egg apparatus in *Arundo formosa* Hack (Poaceae). *International Journal of Plant Sciences* **158**: 713–726.

Jayalakshmi, R. and K.K. Lakshmanan. 1982. Contribution to the embryology of *Spinefex littoreus. J. Madras Univ. Sect.* **B. 42**: 51–62.

Jefferson, R. 1994. Apomixis: a social revolution for agriculture? *Biotechnol. Dev. Monitor.* No. **19**: 14–16.

Joachimiak, A. 1981. Antipodals in *Phleum boehmeri* Wib. : Development and structure. *Acta Biol. Cracov. Ser. Bot.* **23**: 25–35.

Johansen, D.A. 1945. A critical survey of the present status of plant embryology. *Bot. Rev.* **11**: 87–107.

Johansen, D.A. 1950. *Plant Embryology.* Waltham. Mass. U.S.A.

Johnston, G.W. 1953. Observation on twin embryo sacs in *Sorghum vulgare. Phytomorphology.* **3:** 313–315.

Johri, B.M. 1963. Embryology and Taxonomy. In: Maheshwari, P. (ed.). *The Recent Advances in the Embryology of Angiosperms.* International Society of Plant Morophologists, Univ. of Delhi, Delhi, pp. 395–444.

Johri, B.M. and K.B. Ambegaokar. 1975. The antipodals in *Triticale. Phytomorphology.* **25**: 112–117.

Johri, B.M. and K.B. Ambegaokar. 1977. Seed development in *Triticale* — 1. *Phytomorphology* **27**: 190–197.

Johri, B.M., K.B. Ambegaokar and P.S. Srivastava. 1992. *Comparative embryology of Angiosperms.* Springer–Verlag, Berlin.

Jones, J.W. 1928. Polyembryomy in rice. *J. Amer. Sci. Agron.* **20**: 774.

Joppa, L.R., F.H. McNeil and J.R. Walsh. 1966. Pollen and anther development in cytoplasmic male sterile wheat *Triticum aestivum. Crop. Sci.* **6**: 296–97.

Jorgensen, C.A., T.H. Sorensen and M. Westergaard, 1958. The flowering plants of Greenland. A taxonomical and cytological survey. *Biol. Skr.* **9**: 1–172.

Julen, G. 1961. The effect of X–rays on the apomixis of *Poa pratensis*. IAEA Vienna pp. 527–532.

Juliano, J.B. and M.J. Aldama. 1937. Morphology of *Oryza sativa* L. *Philipp. Agric.* **26:** 1–134.

Just, T. 1946. The use of embryological formulas in Plant taxonomy. *Bull. Torrey. Bot. Cl.* **73**: 351–355.

Kalantri, S.R. 1990. Embryology of *Eragrostis tenella* (Linn.) P. Beauv. with a brief discussion on morphology of embryo. 12th all India Bot. Conf. Abstract.

Kalinina, L.V. 1959. Development of embryo in corn grains. *Kukaruza* **6**: 23–24.

Kaltsikes, P.J. 1973. Early seed development in hexaploid *Triticale. Can. J. Bot.* **51**: 2291–2300.

Kaltsikes, P.J. and D.G. Roupakias. 1975. Endosperm abnormalities in *Triticale Secale* combinations. 2. Addition and substitution lines. *Can. J. Bot.* **53**: 2068–2076.

Kaltsikes, P.J., D.G. Roupakias and J.B. Thomas. 1975. Endosperm abnormalities in *Triticum–Secale* combinations. 1. X *Triticosecale* and its parental species. *Can. J. Bot.* **53**: 2050–2067.

am, Y.K. and J. Maze. 1974. Studies on the relationships and Evolution of super–specific taxa utilizing developmental data. II. Relationships and evolution of *Oryzopsis hymenoides, O. virescens, O. kingii, O. micrantha* and *O. asperifolia. Bot. Gaz.* **135**: 227–247.

Kamra, O.P. 1960. Occurrence of binucleate and multinucleate pollen mother cells in *Hordeum. Hereditas.* **46**: 536–542.

Kandelaki, G.V. 1961. Studies of embryogenesis in Georgian wheat hybrids. In: *Plant Morphogenesis.* Moscow **2**: 362–365.

Kandelaki, G.V. 1978. 'Apomixis and spontaneous polyploidy in wheat–rye hybrids. *Proc. XIV Int. Congr. Genet.* Moscow. abstrct. Part II. p. 169.

Kapil, R.N. and A.K. Bhatnagar. 1975. A fresh look at the process of double fertilization in Angiosperms. *Phytomorphology* **25**: 334–368.

Kapil, R.N. and S.C. Tiwari. 1977. Impact of embryology on Agriculture. In: Symposium on Basic Sciences and Agriculture. pp. 43–52. Indian National Science Academy, New Delhi.

Karas, I. and D.D. Cass. 1976. Ultrastructural aspects of sperm cell formation in rye: Evidence for cell plate involvement in generative cell division. *Phytomorphology.* **26**: 36–45.

Kasparyan, A.S. 1938. Haploids and haplo–diploids among hybrid twin seedlings in wheat. *Dokl. Akad. Nuak. S.S.S.R.* **20**: 53–56.

Kellogg, Elizabeth, A. 1987. Apomixis in the *Poa secunda* complex. *Amer. J. Bot.* **74**: 1431–1437.

Kellog, E.A. 1990. Variation and species limits in agamospermous grasses. *Systematic Botany.* **15**: 112–123.

Kempanna, C., A. Seetharam and G. Shivashankar. 1971. Multinucleate pollen mother cells in *Triticale. Curr. Sci.* **40**: 301–302.

Kennedy, P.B. 1900. The structure of the caryopsis of the grasses with reference to their morphology and classification. *U.S. Dept. Agr. Div. Agros. Bull.* **19:** 1–44.

Khan, A.S. 1990. Apomixis in *Urochloa panicoides* Beauv. and *Cymbopogon winterianus* Fowitt. *12th all India Bot. Conf. Abstract.*

Khattra, S. and G. Singh. 1989. Histochemical studies of anther development in male sterile *Pennisetum typhoides. Acta Bot. Indica.* **17**: 159–162.

Khazava, I.I. 1970. The embryology of cultivated species of *Sorghum. Bot. Zhur.* **55**: 93–102.

Khokhlov, S.S. 1976. Apomixis and breeding. Amerind, New Delhi.

Khosla, S. 1946. Development morphology in some Indian millets. *Proc. Indian Acad. Sci.* B. **24**: 207–224.

Khush, G.S. 1994. *Apomixis: Exploiting Hybrid vigour.* International Rice Research Institute, Las Banos, Laguna, Philippines.

Khush, G.S. 1997. Apomixis — A new frontier of exploiting hybrid vigour in Rice. *The Botanica* **47**: 1–13.

Khush, G.S., D.S. Brar, J. Bennett and S.S. Virmani. 1994. Apomixis for rice improvement. In: *Apomixis: Exploiting hybrid vigour in rice*. G.S. Khush (ed.). IRRI, Manila p. 1–24.

Kiellander, C.L. 1935. Apomixis bei *Poa serotina. Bot. Notiser* **35**: 87–95.

Kiellander, C.L. 1937. On the embryological basis of apomixis in *Poa palustris. Svensk Bot. Tidskr.* **31**: 425–429.

Kiellander, C.L. 1941. Studies on apospory in *Poa pratensis*. var. *alphina. Svensk Bot. Tidskr.* **35**: 321–332.

Kiesselbach. T.A. 1926a. False polyembryony in Maize. *Amer. J. Bot.* **13**: 33–34.

Kiesselbach. T.A. 1926b. Fasciated Kernels, reversed kernels and related abnormalities in maize. *Amer. J. Bot.* **13**: 35–39.

Kihara, H. 1936. A diplo–haploid twin plant in *Triticum durum. Agric. & Hort.* Tokyo. **11**: 1425–1434.

Kihara, H. and T. Hori. 1966. The behaviour of nuclei in germinating pollen grains of wheat, rice and maize. *Zuchter* **36**: 145–150.

Kihara, H. and I. Hirayoshi. 1942. Development of pollen grain in rice. *N.E.* **17**: 685–690.

Kindiger, B. and C. Dewald. 1994. Genome accumulation in Eastern gamagrass (*Tripsacum dactyloides* (L.) L. Poaceae. *Genetics* **92**: 197–201.

Kindiger, B. and C. Dewald. 1996. A system for genetic change in apomictic eastern gamagrass. *Crop. Sci.* **36**: 250–255.

Kindiger, B., V. Sokovlov and I.V. Khatypova. 1996. Evaluation of apomictic reproduction in a set of 39 chromosome maizeT— *Tripsacum* back cross hybrids. *Crop Sci.* **36**: 1108–1113.

Kindigr, B., V. Sokovlov and C. Dewald. 1996. A comparison of apomictic reproduction in eastern gama grass (*Tripsacum dactyloides* (L.) L.) and maize — *Tripsacum* hybrids. Genetica 97: 103–110.

Kirpes, C.C., L.G. Clark and N.R. Lersten. 1996. Systematic significance of pollen arrangement in microsporangia of Poaceae and Cyperaceae: review and observations on representative taxa. *Amer. J. Bot.* **83**: 1609–1622.

Kljucareva, M. 1958. New facts on the fertilization of barley. *Dokl. Akad. Nauk SSSR* **123**: 1128–1130.

Klyuchareva, M.V. 1983. Extrusion of nuclear material in proembryos of gramineous plants. *Dokl. Bot. Sci. Proc. Acad. Sci.* USSR 268–270; 19–21 (in Russian).

Knox, R.B. 1967. Apomixis: Seasonal and population differences in a grass. *Science* **157**: 325–326.

Knox, R.B. and J. Heslop–Harrison. 1963. Experimental control of aposporous apomixis in a grass of the andropogoneae. *Bot. Notiser.* **116**: 127–141.

Kojima, A., T. Kojono, Y. Nagato and K. Hinata. 1994. Non–parthenogenetic plants detected in Chinese chive, a facultative apomict. *Breeding Sci.* **44**: 143–149.

Koltunow, A.M. 1993. Apomixis: embryo sacs and embryos formed without meiosis or fertilization in ovules. *The Plant Cell* **5**: 1425–1437.

Komuro, H. 1922. A polyembryonal plant of *Oryza sativa*. *Bot. Mag.* (Tokyo) **11**: 23–24.

Konstantinov, A.V. 1963. Contribution to the cyto–embryology of wheat, rye, amphidiploids. *Trud. Prikl. Bot. Genet. Selek.* **35**: 140–145.

Korobova, S.N. 1961. Microsporogenesis and development of the pollen grain in corn. *Dokl. Akad. Nauk. SSSR* **136:** 3–9.

Koroboa, S.N. 1959. On the course of fertilization of barley. *Dokl. Nauk SSSR* **127**: 921–923.

Koroboa, S.N. 1962. Embryology of maize. 294–314. In: Alexandrova, V.G. (ed.) *Morphology and Anatomy of Plants.* Moscow, USSR.

Korobova, S.N. 1971. On the development of the specific cells during micro– and macrosporangium formation in *Zea mays*. *Ann. Univ. ARERS* **9:** 58–61 (in Russain).

Kostoff, D. 1939. Frequency of polyembryomy and chlorophyll deficiency in rye. *Dokl. Akad. Nauk SSSR.* **24**: 479–482.

Koul, A.K. 1959. Antipodals during the development of caryopsis in *Euchlaena mexicana*. *Agra. Univ. J. Res. (Sci.).* **8**: 31–33.

Koul, A.K. 1970a. Cytoembryological studies in oriental Maydeae I: *Coix aquatica* Roxb. *Proc. Nat. Acad.* Sci. **40**: 163–178.

Koul, A.K. 1970b. Cytoembryological studies in oriental Maydeae II: *Chionachne koenigii. Proc. Nat. Acad. Sci.* **40**: 178–190.

Koul, A.K. and B.G. Sahi. 1962. Morphological studies on *Euchlaena mexicana*. *Agra Univ. J. Res. (Sci.)* **11**: 195–207.

Krishnaswamy, N. 1939. Cytological studies in a haploid plant of *Triticum vulgare. Hereditas*. **25**: 77–86.

Krishnaswamy, N. and G.N.R. Ayyangar. 1930. Polyembryony in *Eleusine coracana* Gaertn. *Madras Agric. J.* **18**: 593–595.

Krishnaswamy, N. and G.N.R. Ayyangar. 1937. Cytological studies in *Eleusine coracana. Beih. Bot. Ztbl.* **57:** 297–318.

Krishnaswamy, N. and G.N.R. Ayyangar. 1941. An autotriploid in the pearl millet *Pennisetum typhoides* Stapf & Hubbard. *Proc. Indian Acad. Sci.*. **B13**: 9–33.

Kruse, A. 1969. Intergeneric hybrids between *Triticum aestivum* L. (V. Koga II, 2n = 42) and *Avena sativa* L. (V. Stal. 2n = 42) with pseudogamous seed formation. *Roy. Vet. Agric. Univ. Yearbook.* Copenhagen, pp. 188–200.

Kubien, E. 1968. Cytological processes during the development of antipodals in *Ammophila arenaria* Link. *Acta Biol. Cracow. Ser. Bot.* **11**: 21–29.

Kulkarni, A.R. 1981. Embryology of *Eragrosis cilianensis. Proc. 68th Indian Sci. Congr. Part III. Abst.* p. 53.

Kulkarni, A.R. 1982. Embryology of *Andropogon pumilus. Proc. 69th Indian Sci. Congr. Part III. Abst.* p. 88.

Kulkarni, A.R. and V.R. Dnyansagar. 1984. Embryology of *Eragrostis ciliensis. J. Indian Bot. Soc.* **63**: 206–213.

Kumar, L.S.S. 1942. Non hertitable polyembryony in *Andropogon sorghum. Curr. Sci.* **11**: 242.

Kurup, S. 1973. Cytoembryological studies in *Pennisetum hohenackeri* Hochst. *Journ. Mysore Univ.* **25**: 1–7.

Kubien, E. 1968. Cytological processes during the development of antipodals in *Ammophila arenaria* Link. *Acta Biol. Cracow Ser. Bot.* **11**: 21–29.

Kuwada, Y. 1910a. On the development of the pollen and the embryo sac and the formation of endosperm of *Oryza sativa. Bot. Mag. Tokyo.* **23**: 268–280.

Kuwada, Y. 1910b. A cytological study of *Oryza sativa. Bot. Mag. Tokyo* **24**: 267–281.

Kuwada, Y. 1911. Meiosis in the pollen mother cells of *Zea mays. Bot. Mag. Tokyo.* **25**: 163–181.

Laikova, L.I. 1984. Cytoembryological study of maize. X. *Tripsacum* hybrids. p. 79–87. In: D.F. Petrov (ed.) *Apomixis and its role in evolution and breeding*. Nauka, USSR.

Lakshmanan, K.K. 1972. The monocot embryo. In: Verghese T.M. & Grover, R.K. (ed.) *Vistas in Plant Scineces*. Vol. **2**. *Intl. Bio. Sci. Hissar.* pp. 61–110.

Lakshmanan, K.K. and K.B. Ambegaokar. 1984. Polyembryony. In: B.M. Johri (ed.) *Embryology of Angiosperms*. Springer Verlag, Berlin.

Lakshmanan, K.K. and R. Jayalakshmi. 1980. Preliminary observation on the embryology of *Spinifex littoreus. Curr. Sci.* **49**: 325–326.

Lawrence, G.H.M. 1951. *Taxonomy of vascular plants.* Macmillan.

Lebegue, A. 1952. La polyembryonie chez les Angiospermes. *Bull. Soc. Bot. France.* **99**: 329–367.

Leblanc, O., M. Duemas, M. Hernandez, S. Bello, V. Garcia, J. Berth and Y. Savidan. 1995. Chromosome doubling in *Tripsacum.* The production of artificial sexual tetraploid plants. *Plant Breeding.* **114**: 226–230.

Leblanc, O., D. Grimanelli, D. Gonzalez–de–Leon and Y. Savidan. 1995. Detection of the apomictic mode of reproduction in maize – *Tripsacum* hybrids using maize RFLP markers. *Theor. Appl. Genet.* ***90***: 1198–1203.

Leblanc, O., D. Grimandli, N. Islam-Faridi, J. Berthand and Y. Savidan. 1996. Reproductive behaviour in maize — *Tripsacum* polyploid plants. Implications for the transfer of apomixis into maize. *Journ. Heredity* 87: 108–111.

Leblanc, O., M.D. Peel, J.G. Carman and Y. Savidan. 1995. Megasporogenesis and mega gametogenesis in several *Tripsacum* species (Poaceae) *Amer. Journal Bot.* **82**: 57–63.

Leblanc, O. and Y. Savidan. 1994. Timing of megasporogenesis in *Tripsacum* species (Poaceae) as related to the control of apomixis and sexuality. *Polish Bot. Studies* **8**: 75–81.

Liebenberg, H. 1961. Apomiksie by *Eragrostis* en *Themeda.* Ongepuliseerde M.Sc. Verhandeling. Universitat Van Stellenbosch.

Liebenberg, H. 1990. Cytotaxonomic studies in *Themeda triandra* Forssk. Part III. Sexual and apomictic embryo sac development in 53 Collections. *South Afr. J. Bot.* **56**: 554–559.

Liebenberg, H. and R. Dev. Pienaar. 1962. Apomiksie by *Eragrostis* en *Themeda* verrigtinge Van die 2de Kongres Van die Suid–Afrikaanse Genetiese Vereniging. 150–153.

Libenberg. H., J.Lubbinge and A. Fossey. 1993. Cytotaxonomic studies in *Themeda triandra* Forssk. 4. A population cytogenetic study of a contact zone between tetraploid and hexaploid population. *South Afr. J. Bot.* **59:** 305–310.

Linder, H.P. 1987. The evolutionary history of Poales/Restionales — A hypothesis. *Kew Bull.* **42**: 297–318.

Linder, H.P. and I.K. Ferguson. 1985. Notes on the pollen morphology and phylogeny of the Restionales and Poales. *Grana* **29:** 65–76.

Lindley, J. 1853. *The Vegetable Kingdom* (ed. 3) London.

Liu, Z.W., R.R.C. Wang and J.G. Carman. 1995. Hybrids and back cross progenies between wheat (*Triticum aestivum*) and apomictic Australian wheat grass (*Elymus rectisetus*) Karyotypic and genomic analysis. *Theor. Appl. Genet.* **89**: 599–605.

Longley, B., T. Rabau and B.P. Louant. 1985. Development floral chez *Eragrostis tef.* dynamique des gameto phytogensis. *Can. J. Bot.* **63**: 1900–1906.

Lotsy, J.P. 1911. Vortrage uber botanisetie stammes ges–chickte. *Cormophyta Siphenegamia.* Band 3. G. Fischer Jena.

Lubbers, E.L., L. Arthur, W.W. Hanna and P. Ozias–Akins. 1994. Molecular markers shared by diverse apomictic *Pennisetum* species. *Theor. Appl. Genet.* **89**: 636–642.

Lutts, S., J. Ndikumana and B.P. Louant. 1991. Fertility of *Brachiaria ruziziensis* in interspecific crosses with *Brachiaria decumbens* and *Brachiaria brizantha*: Meiotic behaviour, pollen viability and seed set. *Euphytica* **57**: 267–274.

Lutts, S., J. Ndikumana and B.P. Louant. 1994. Male and female sporogenesis in apomictic *Brachiaria brizantha, Brachiaria decumbens* and F_1 hybrids with sexual colchicine induced tetraploid *Brachiaria ruziziensis. Euphytica* **78**: 19–25.

Luxova, M. 1967. Fertilization of Barley (*Hordeum distichum* L.). *Biol. Plant.* (Praha), **9:** 301–307.

Luxova, M. 1968. The temporal courses of fertilisation and localisation of starch in the barley pistil. *Biologia Pl.* **10**: 10–14.

Luxova, M. and O. Erdelska. 1965. Plasmatic inclusions in fusing barley polar nuclei. *Biologia (Bratislava)* **20:** 890–893.

Macleod, A.M. and H. McCorquodale. 1958. Comparative studies of embryo and endosperm in the Gramineae. *J. Inst. Brew.* **64**: 162–170.

Mahalingappa, M.S. 1976. *Embryological studies in some members of Gramineae.* Ph.D. thesis, Karnataka University, Dharwar.

Mahalingappa, S. 1977. Gametophytes of *Eleusine compressa. Phytomorphology* **27**: 231–239.

Mahalingappa, S. 1978. Microsporogenesis and male gametophyte in *Cynodon dactylon. Curr. Sci.* **47**: 594–495.

Maheshwari, P. 1950. *An Introduction to the embryology of Angiosperms.* New York.

Maheshwari, P. 1963. *Recent advances in the embryology of Angiosperms.* Delhi.

Maheshwari, P. 1964. Embryology in relation to taxonomy. *Int. ser. Monogr. Pure & Appl. Biol.* **7:** 55–97.

Maheshwari, S.C., N. Maheswari, J.P. Khurana and S.K. Sopory. 1998. Engineering apomixis in crops: A challenge for Plant Molecular Biologists in the next century. *Curr. Sci.* **75**: 1141–1147.

Mangelsdorf, P.C. and Reeves. 1939. The origin of Indian corn and its relatives. *Texas Agric. Exp. Sta. Bull.* No. **574**.

Manning, J.C. and H.P. Linder. 1990. Cladistic analysis of patterns of endothecial thickenings in the Poales/Restionales. *Amer. J. Bot.* **77**: 196–210.

Mares, D.J., K. Norstog and B.A. Stone. 1975. Early stages in development of wheat endosperm. The change from free nuclear to cellular endosperm. *Aust. J. Bot.* **23**: 311–326.

Mares, D.J., B.A. Stone, C. Jeffrey and K. Norstog. 1977. Early stages in the development of wheat endosperm. II. Ultrastructural observations on cell wall formation. *Aust. J. Bot.* **25**: 599–613.

Marshall, D.R. and R.W. Daines. 1977. *Euphytica* **26**: 661.

Martin, F.W. 1970. Compounds of the stigmatic surface of *Zea mays* L. *Ann. Bot.* **34**: 835–842.

Martinez, E.J., F. Espinoza and C.L. Quarin. 1994. B III progeny (2n + n) from apomictic *Paspalum notatum* obtained through early pollination. *J. Hered.* **85**: 295–297.

Matzk. 1991. New efforts to overcome apomixis in *Poa pratensis* L. *Euphytica* **55**: 65–72.

Maze, J. and L.R. Bohm. 1973. Comparative embryology of *Stipa elmeri* (Graminae). *Can. J. Bot.* **51**: 235–247.

Maze, J. and L.R. Bohm. 1974. Embryology of *Agrostis interrupta* (Graminae). *Can. J. Bot.* **52**: 365–379.

Maze, J. and L.R. Bohm. 1977. Embryology of *Festuca microstachys* (Graminae). *Can. J. Bot.* **55**: 1768–1782.

Maze, J., L.R. Bohm and C.E. Beil. 1972. Studies on the relationships and evolution of superspecific taxa utilizing developmental data. *Stipa lemmonnii* (Gramineae). *Canad. J. Bot.* **50**: 2327–2352.

Maze, J., L.R. Bohm and L.E. Mehlenbacher. 1970. Embryo sac and early ovule development in *Oryzopsis miliacea* and *Stipa tortilis*. *Can. J. Bot.* **48**: 27–41.

Maze, J., N.G. Dengler and L.R. Bohm. 1971. Comparative floret development in *Stipa tortilis* and *Oryzopsis miliacea*. *Bot. Gaz.* **132:** 273–298.

Maze, J. and S.C. Lin. 1975. A study of the mature megagametophyte of *Stipa elmeri*. *Can. J. Bot.* **53**: 2958–2977.

Mazzucato, A. 1995a. Italian germplasm of *Poa pratensis* L. I. Variability and mode of reproduction. *Journal of Genetics and Breeding* **49**: 111–118.

Mazzucato, A. 1995b. Italian germ-plasm of *Poa pratensis* II. Isozyme progeny test to characterize genotypes for their mode of reproduction. *Journal of Genetics and Breeding* **49**: 119–125.

Mazzucato, A., P. Anton, M. Den Nijis and M. Falcinelli. 1996. Estimation of Parthenogenesis frequency in Kentucky blue grass with auxin-induced parthenocarpic fruits. *Crop Sci.* **36**: 9–16.

Mazzucato, A., G. Barcaccia, M. Pezzotti and M. Falcinelli. 1995. Biochemical and molecular markers for investigating the mode of reproduction in the facultative apomict *Poa protensis* L. *Sexual Plant Reproduction.* **8:** 133–138.

Mazzucato, A., M. Falcinelli and Fabio Veronesi. 1996. Evolution and adaptedness in a facultatively apomictic grass, *Poa ratensis*. *Euphytica* **92:** 13–19.

McClintock, B. 1977. The clonal basis of development. In: S. Subtelny & I.M. Sussex (eds.) *Symp. Soc. Dev. Biol.* **36**: 217–246.

McWilliam, J.R., K. Shankar and R.B. Knox. 1970. Effects of temperature and photo–period on growth and reproductive development in *Hyparrhenia hirta*. *Australian J. Agric. Res.* **21**: 557–569.

Mehlenbacher, L.E. 1970. Floret development, embryology and systematic position of *Oryzopsis hendersonii* (Gramineae). *Can. J. Bot.* **48**: 141–1758.

Melchoir, H. 1964. Engler's *Syllabus der Pflanzenfamilien.* (Revised ed.) Berlin.

Mello, L.V., W.J. Silva, H.P. Medina Filho and R. Balvve. 1995. Breeding systems in *Coix lacryma–jobi* populations. *Euphytica* **81**: 217–221.

Mengesha, M. and A.T. Guard. 1966. Development of the embryo of teff, *Erasrostis tef. Can. J. Bot.* **44**: 1971–1975.

Merry, J. 1941. Studies on the embryology of *Hordeum sativum.* 1. The development of embryo. *Bull. Torrey Bot. Club.* **68**: 585–598.

Merwe, R.B. Van der. 1957. An embryological study on *Themeda triandra. J. South Afri. Bot.* **23**: 139–149.

Miller, E.C. 1920. Development of the pistillate spikelet and fertilization in *Zea mays. J. Agri. Res.* **18**: 255–267.

Miles, J.M. and M.L. Escandon. 1997. Further evidence on the inheritance of reproductive mode in *Brachiaria. Canad. J. Pl. Sci.* **77**: 105–107.

Misra, K.C. 1948. Polyembryony in *Dicanthium annulatum. Curr. Sci.* **17:** 91.

Modilevski, J.S. and R.A. Beilis. 1937. Zur embryologie und zytologie Von Weizen. I. Embryogenesis des Weizen. Vomarchesporium biszum embryo. *Zhur. Inst. Bot. Akad. Nauk URSR* **21/22**: 127–141.

Modilevski, J.S. and R.A. Beilis. 1938. On the embryology and cytology of the wheat plant. II. The stages of maturing of the embryo and caryopsis, of their germination and earing. *Zhurn. Inst. Bot. Akad. Nauk. URSR* **26/27**: 13–39.

Mogensen, H.L. 1982. Double fertilization in Barley and the cytological exploration for haploid embryo formation, embryoless caryopsis and ovule abortion. *Carlesberg Res. Commun.* **47**: 313–354.

Mogensen, H.L. 1984. Quantitative observations on the pattern of synergid degeneration in barley. *Amer. J. Bot.* **71**: 1448–1451.

Mogie, M. 1988. A model for the evolution and control of generative apomixis. *Biol. J. Linn. Soc.* **35**: 127–153.

Mogie, M. 1992. *The evolution of a sexual reproduction in plants.* Chapman and Hall, London.

Mohamed, A.H. and Gould, F.W. 1966. Bio–systematic studies in the *Bouteloua curtipendula* V. Megasporogenesis and embryo sac development. *Amer. J. Bot.* **53**: 166–169.

Mohan Ram, H.Y. and B. Hari Gopal. 1981. Some observartions on the flowering of bamboos in Mizoram. *Curr. Sci.* **50**: 708–710.

Mol, R., E. Matthys–rochon and C. Dumas. 1994. The kinetics of cytological events during double fertilization in *Zea mays* L. *Plant J.* **5**: 197–206.

Morgan, R.N., J. Alvernaj, L. Arthur, W.W. Hanna and P. Ozias–Akins. 1997. Genetic characterization and floral development of female sterile and stubby head, two aposporous mutants of pearl millet. *Sexual Plant Reproduction* **10**: 127–135.

Morgan, R.N., P. Ozias–Akins and W.W. Hanna. 1998. Seed set in Apomictic BC_3 Pearl millet. *Int. J. Plant. Sci.* **159:** 89–97.

Morrison, J.W. 1955. Fertilisation and post–fertilisation development in wheat. *Can. J. Bot.* **33**: 168–176.

Morrison, I.N. and T.P. O'Brien. 1976. Cytokinesis in the developing wheat grain division with and without a phragmoplast. *Planta* **130**: 57–67.

Moskova, R. and S. Mandlikova. 1970. Cytoembryological and Karyological studies in *Panicum miliaceum* L. *C.V. Miranovsho. Gent. Plant Breed.* **3:** 485–491.

Moskova, R.D. and M.S. Yakovlev. 1974. Apomixis of *Bothriochloa ischaemum* L. flowering, pollination, fertilization, In: Linskens, H.F. (ed.) Fertilization in higher plants. North–Holland Publ. Amsterdam. pp. 341–349.

Mowery, M. 1929. Development of the pollen grain and the embryo sac of *Agropyron repens. Bull. Torrey Bot. Cl.* **56**: 319–324.

Mujeeb Kazi, A. 1981. Apomictic progeny derived from intergeneric *Hordeum–Triticum* hybrids. *J. Hered.* **72**: 284–285.

Mujeeb Kazi, A. 1996. Apomixis in trigeneric hybrids of *Triticum aestivum / Leymus racemosus / Thinopyrum elongatum. Cytologia* **61**: 15–18.

Muniyamma, M. 1976. Cytoembryological study of *Agrostis pilosulu. Can. J. Bot.* **54**: 2490–2496.

Muniyamma, M. 1976. *Cytoembryological studies in Gramineae*. Ph.D. Thesis, Univ. Mysore.

Muniyamma, M. 1977. Triploid embryos from endosperm *in vivo. Ann. Bot.* **10:** 1077–1079.

Muniyamma, M. 1978. Variations in Microsporogenesis and the development of embryo sacs in *Echinochloa stagnina* (Gramineae). *Bot. Gaz.* **139**: 87–94.

Muntzing, A. 1933. Apomictic and sexual seed formation in *Poa. Hereditas.* **17**: 131–154.

Muntzing, A. 1937. Polyploidy from twin seedlings. *Cytologia* Fuji Jubl. Vol. pp. 211–227.

Muntzing, A. 1938. Note on heteroploid twin plants from eleven genera. *Hereditas* **24**: 487–491.

Muntzing, A. 1940. Further studies on apomixis and sexuality in *Poa. Heridatas* **26**: 115–140.

Muntzing, A. 1943. Characteristics of two haploid twıns in *Dactylis glomerata. Hereditas* **29**: 134–140.

Muntzing, A. 1966. Apomixis and sexuality in new material of *Poa alpina* from middle Sweden. *Hereditas* **54**: 314–337.

Murthy, U.R. 1973. Polyploidy and apomixis in *Apluda mutica var. aristata* (L.) Pilger. *Cytologia* **38**: 347–356.

Murthy, U.R. 1993. Appraisal on the presentation of research on apomixis in sorghum. *Curr. Sci.* **64**: 315–318.

Murthy, U.R. and N.G.P. Rao. 1972. Apomixis in breeding grain *Sorghum.* In: N.G.P. Rao and L.R. House (ed.) *Sorghum in Seventies.* Oxford & I.B.H. New Delhi.

Murthy, U.R., K.F. Schertz and E.C. Bashaw. 1979. Apomictic and sexual reproduction in *Sorghum. Indian J. Genet. & Plant Breed.* **39**: 271–278.

Nagendran, C.R. and M.S. Dinesh. 1988. *The Embryology of Angiosperms: a classified Bibliography (1965–1985).* Indira Publishing House, Michigan.

Nagato, Y. 1981. Embryological studies on the evolutionary path in Asian rice. *Bot. Gaz.* **142**: 274–278.

Nakagawa, H. 1990. Embryo sac analysis and crossing procedure for breeding apomictic Guinegrass (*Panicum maximum* Jacq.) *Jarq.* **24**: 163–168.

Nakagawa, H. 1993. Manipulation of apomixis in guinea grass (*Panicum maximum* Jacq). *Proc. XV Intern. Bot. Cong.* 184

Nakagawa, H., Shimizu and W.W. Hanna. 1993. Cytology of "Natsukaze" guinea grass, a natural apomictic hybrid between a sexual and an apomictic plant. *Journal of Japanese Society of Grassland Science* **39**: 374–380.

Namirawa, S. and J. Kawakami. 1934. On the occurrence of haploid, triploid and tetraploid plants in twin seedlings of common wheat. *Proc. Imp. Acad.* (Tokyo). **10**: 668–671.

Narasa Reddy, R., L.L. Narayana and N.G.P. Rao. 1979. Apomixis and its utilisation in grain sorghum. II. Embryology of F_2 progeny of reciprocal crosses between R 473 and 302. *Indian Acad. Sci.* **88**: 445–450.

Narayan, K.N. 1951. *Cytogenetic studies in apomictic Pennisetum species.* Ph.D. Thesis, Univ. of California, Berkeley.

Narayan, K.N. 1955a. Cytogenetic studies of apomixis in *Pennisetum: Pennisetum clandestinum* Hochst. *Proc. Indian Acad. Sci.* **B.41**: 196–208.

Narayan, K.N. 1955b. Cytogenetic studies of apomixis in *Pennistetum.* II. *Pennisetum latifolium* Spreng. *J. Mysore Univ.* **14B**: 33–42.

Narayan, K.N. 1955c. Apomixis in Angiosperms. *J. Mysore Univ.* **14**: 411–415.

Narayan, K.N. 1962. Apomixis in some species of *Pennisetum* and in *Panicum antidotale.* In: *Plant Embryology* (Ed.) P. Maheshwari, CSIR, New Delhi, pp. 55–61.

Narayan, K.N. 1967–68. Cytogenetical studies in Gramineae and Compositae. *J. Mysore Univ. (N.S.) Sect. B.* Golden Jubilee Vol. p. 144–156.

Narayana, P.S. 1986. Reproductive behaviour of VZM 2B: A male sterile sorghum line. *J. Indian Bot. Soc.* **65**: 131–135.

Narayanaswami, S. 1940. Megasporogenesis and origin of triploids in *Saccharum. Indian J. Agric. Sci.* **10**: 534–551.

Narayanaswami, S. 1952. Microsporogenesis and male gametophyte in *Eleusine coracana. Curr. Sci.* **21**: 19–21.

Narayanaswami, S. 1953. The structure and development of caryopsis in some Indian millets. 1. *Pennisetum typhoideum. Phytomorophology* **3:** 98–112.

Narayanaswami, S. 1954. The structure and development of the caryopsis in some Indian millets 2. *Paspalum scrobiculatum. Bull. Torrey Bot. Cl.* **81**: 288–299.

Narayanaswami, S. 1955a. The structure and development of the caryopsis in some Indian millets 3. *Panicum miliare* and *P. miliaceum. Lloydia* **18**: 61–73.

Narayanaswami, S. 1955b. The structure and development of the caryopsis in some Indian millets 4. *Echinochloa frumentacea. Phytomorphology* **5:** 161–171.

Narayanaswami, S. 1955c. The structure and development of the caryopsis in some Indian millets 5. *Eleusine coracana. Michigan Acad. Sci. Arts Letters.* **40:** 33–46.

Narayanaswami, S. 1956. Structure and development of the caryopsis in some Indian millets 6. *Setaria italica. Bot. Gaz.* **118**: 112–122.

Natesh, S. and M.A. Rau. 1984. The Embryo. In: Johri, B.M. (ed.) *Embryology of Angiosperms.* Springer–Verlag. Berlin.

Naumova, T.N. 1993. *Apomixis in Angiosperms: Nucellar and integumentary embryony.* CRC Press, Boca Raton.

Naumova, T.N., A.P.M. Den Nijs and M.T.M. Willemse. 1993. Quantitative analysis of aposporous parthenogenesis in *Poa pratensis* genotypes. *Acta Bot. Neerl.* **42**: 299–312.

Naumova, T.N. and M.T.M. Willemse. 1995. Ultra structural characteristics of apospory in *Panicum maximum. Sexual Plant Reproduction.* **8**: 197–204.

Naumova, T.N. 1997. Apomixis in Tropical fodder crops: Cytological and functional aspects. *Euphytica* **96**: 93–99.

Newffer, M.G. 1964. Tetrasporic embryo sac formation in trisomic sectors of maize. *Science* **144**: 874–876.

Nielsen, E.L. 1946. The origin of the multiple macrogametophytes in *Poa pratensis. Bot. Gaz.* **108:** 41–50.

Nielsen, E.L. 1947a. Developmental sequence of embryo and endosperm in apomictic and sexual forms of *Poa pratensis. Bot. Gaz.* **108:** 531–534.

Nielsen, E.L. 1947b. Macrosporogenesis and fertilization in *Bromus inermis. Amer. J. Bot.* **34:** 431–433.

Nikolaevskaya, T.S. 1974. Development of *Dactylis glomerata* Caryopsis. *Bot. Zh.* **59**: 1623–1630.

Nirmala, A., B.H. Rao and P.N. Rao. 1993. Variable sterility and apospory in pearl millet. *Indian J. Genet.* **53**: 47–54.

Nishimura, M. 1922a. On the germination and the polyembryony of *Poa pratensis. Bot. Mag.* (Tokyo) **36**: 47–54.

Nishimura, M. 1922b. Comparative morphology and development of *Poa pratensis*, *Phleum pratense* and *Setaria italica. Jap. J. Bot.* **1**: 55–85.

Nishimura, M. 1922c. Comparative morphology and development of *Poa pratensis. Bot. Mag. Tokyo.* **36**:

Nishiyama, I. 1970. Male sterility caused by cooling treatment at the young microspore stage in rice plants. Part 6. Electron microscopical observations on normal tapetal cells at the critical stage. *Proc. Crop. Sci. Soc. Jap.* **39**: 474–479.

Nishiyama, I. 1976. Male sterility caused by cooling treatment at the young microspore stage in rice plants. Part 13. Ultrastructure of tapetal hypertrophy without primary wall. *Proc. Crop. Sci. Soc. Jap.* **45**: 270–278.

Nishiyama, I. and T. Yabuno. 1979. Triple fusion of the primary endosperm nucleus as a cause of interspecific incompatibility in Avena. *Euphytica* **28**: 57–65.

Nissen, O. 1937. Spalteapringenes storrelse has tvillingplanter med ulike Kromosomtall. *Bot. Notiser.* **1937:** 28–34.

Nitzsche, W. 1981. Interspecific hybrids between apomictic forms of *Poa palustris* L. x *Poa pratensis* L. Proc. XIV Intl. Grassland Congr. Lexington Ky pp. 155–157.

Nogler, G.A. 1984. Gametophytic Apomixis. In: Johri, B.M. (ed.) *Embryology of Angiosperms.* Springer Verlag, Berlin, pp. 475–518.

Noguchi, Y. 1929. Zur Kenntnis der Befruchtung und Kornildung beider Reispflanze. *Jap. Journ. Bot.* **4**: 385–403.

Norner, C. 1881. Beitrage zur Embryoentwicklung der Graminees. *Flora* **16**: 241–251; 257–266; 273–284.

Norrmann, G.A. 1981. Cytologia y metodo de reproduction en dos especies de *Paspalum* (Gramineae). *Bonplandia* **5:** 149–158

Norrmann, G.A., O.A. Bovo and C.L. Quarin. 1994. Post zygotic seed abortion in sexual diploid x apomictic tetraploid interspecific *Paspalum* crosses. Austral. J. Bot. **42**: 449–456.

Norrmann, G.A., C.L. Quarin and B.L. Burson. 1989. Cytogenetics and reproductive behaviour of different chromosome races in six *Paspalum* species. *Journal of Heredity* **80**: 24–28.

Norstog, K. 1963. Apomixis and polyembryony in *Hierchloe odorata. Amer. J. Bot.* **50:** 815–821.

Norstog, K. 1972. Early development of the barley embryo: Fine structure. *Amer. J. Bot.* **59**: 123–132.

Norstog, K. 1974. Nucellus during early embryogeny in Barley: Fine Structure. *Bot. Gaz.* **135**: 97–103.

Nygren, A. 1946. The genesis of some Scandinavian species of *Calamogrostis. Hereditas* **32**: 131–262.

Nygren, A. 1948a. Further studies in spontaneous and synthetic *Calamogrostis purpurea. Hereditas* **34**: 113–134.

Nygren, A. 1948b. Some interspecific crosses in *Calamogrostis* and their evolutionary consequences. *Hereditas* **34**: 387–413.

Nygren, A. 1949. Apomictic and sexual reproduction in *Calamagrostis purpurea. Hereditas* **35**: 285–300.

Nygren, A. 1950a. A preliminary note on cytological and embryological studies in arctic *Poae. Hereditas* **36**: 231–232.

Nygren, A. 1950b. Cytological and embryological studies in arctic *Poae. Symb. Bot. Upsaliensis* **10:** 1–64.

Nygren, A. 1951. Embryology of *Poa.* Carnegie Inst. Washington Year book **50**: 113–115.

Nygren, A. 1954a. Investigations on North American *Calamogrostis* I. *Hereditas* **40**: 377–397.

Nygren, A. 1954b. Apomixis in the Angiosperms. *Bot. Rev.* **20**: 577–649.

Nygren, A. 1967. Apomixis in Angiosperms. (ed. H.F. Linskens); *Encycl. Pl. Physiol.* **18**: 551–596; Springer–Verlag, Berlin.

Nygren, A. 1962. Artificial and natural hybridization in Europaean *Calamogrostis. Symb. Bot. Upsaliensis.* **17**: 1–105.

Ogorodnikova, V.F. 1986. The genesis and ultrastructure of the sporopollenin wall of tapetal cells in grasses. *Bot. Zhurn. SSSR* **71**: 1366–1371.

Overman, M.A. and H.E. Warmke. 1972. Cytoplasmic male sterility in sorghum II. Tapetal behaviour in fertile and sterile anthers. *J. Hered.* **63**: 226–234.

Ozias–Akins, P., E.L. Lubbers, W.W. Hanna and J.W. McNay. 1993. Transmission of the apomictic mode of reproduction in *Pennisetum:* coinheritance of the trait and molecular markers. *Theor. Appl. Genet.* **85**: 632–638.

Ozias–Akins, P., E.L. Lubbers, L. Arthur and W.W. Hanna. 1994a. Molecular markers for apomixis in *Pennisetum.* In: *Apomixis : Exploiting hybrid vigor in rice.* G.S. Khush (ed.) IRRI, Manila. p. 51–53.

Ozias–Akins, P., E.L. Lubbers, L. Arthur and W.W. Hanna. 1994b. Molecular markers linked with apomixis: Evolutionary and breeding implications. In: *Use of molecular markers in Sorghum and pearl millet breeding for developing countries.* J.R. Witcombe & R.R.Duncan (eds.). ODA, London.

Ozias-Akins, P.., D. Roche and W.W. Hanna. 1998. Tight clustering and hemizygosity of apomixis-linked molecular markers in *Pennisetum squammulatum* implies genetic control of apospory by a divergent locus that have no allelic form in sexual genotypes. *Proc. Natl. Acad. Sci. U.S. America* **95**: 5127–5132.

Pacini, E. 1990. Tapetum and microspore function. In: S. Blackmore & R.B. Knox (eds.) *Microspores, Evolution and Ontogeny.* Academic Press, London.

Pacini, E., P.E. Taylor, M.B. Singh and R.B. Knox. 1992. Development of plastids in pollen and tapetum of rye–grass, *Lolium perenne. Ann. Bot.* **70**: 179–188.

Palser, B.P. 1975. The base of Angiosperm Phylogeny — Embryology. *App. Missouri Bot. Gard.* **62**: 621–646.

Pant, D.D., D.D. Nautiyal and S.K. Chaturvedi. 1982. Insect pollination in some Indian Glumiflorae. *Beitr. Biol. Pflanzen.* **57**: 229–236.

Paul, A.K. and R.M. Datta. 1950. Development of the female gametophyte of *Oryza coarctata.* Sci. and *Cult.* **15**: 487–488.

Paul, A.K. and R.M. Datta. 1953. On the morphology and development of the female gametophyte in *Oryza coarctata. Phillipp. J. Sci.* **62**: 15–18.

Peacock, J. 1993. Genetic engineering and mutagenesis for apomixis in rice. In: K.J. Wilson (ed.) *Proceedings of the International workshop on apomixis in rice*. Rockfellor Foundation, New York, p. 11–12.

Peel, M.A. 1993. Meiocyte callose in aposporic and diplosporic grasses and in hybrids between bread wheat and *Elymus retisetus*. M.S. Thesis. Utah State Univ.

Peel, M.D., J.G. Carman and O. Leblanc. 1997b. Megasporocyte callose in apomicitic buffel–grass, Kentucky Bluegrass, *Pennisetum squammulatum* Fresen, *Tripsacum* L. and Weeping Lovegrass. *Crop Sci.* **37**: 724–732.

Peel, M.D., J.G. Carman, Z.W. Liu and R.R.C. Wang. 1997a. Meiotic anomalies in hybrids between wheat and apomictic *Elymus retisetus* (Nees in Lehm) A. Love & Connor. *Crop Sci.* **37**: 717–723.

Percival, J. 1921. *The Wheat Plant*. London.

Persidsky, D.J. 1940. Embryological and cytological investigations of barley, *Hordeum distichum* L. *Bot. Zhurn. URSR* **1**: 145–153.

Pernes, J. 1971. Etude du mode de reproduction apomixis facultative, du Point deVue de la genetique des populations. Travaux et Documents de PORSTOM. Paris. 9. 66 p.

Pernes, J. 1972. Organisatin evolutive d'un groupe agamique; la section des Maximae du genra *Panicum* (Gramineae). Thesis. Univ. Paris, France. Memoirs ORSTOM **75**: 108 p.

Pernes, J., D. Combes, R. Rene–Chaume and Y. Savidan. 1975. Biologie des populations naturells du *Panicum maximum* Jacq. *Cah* ORSTOM Ser. *Bio.* **10**: 77–89.

Pernes, J., Y. Savidan and R. Rene–Chaume. 1965. *Panicum*: structures genetiques du complexe des "*Maxmae*" et organisation de ses populations naturelles en relation avec la speciation. *Boissiera* **24**: 383–402.

Petrov, D.F., N.I. Belousova, L.I. Likova and R.M. Yatsenko. 1973. First case of transmitting an element of apomixis from *Tripsacum* to corn. *Dokl. Akad Nauk SSSR* 208: 222–224.

Petrov, D.F., N.I. Belosona and E.S. Fokina. 1978. The inheritance pattern of apomictic elements in interspecific hybrids between maize and *Tripsacum*. Abstr. Proc. XIV Int. Congr. Genet. Moscow. Vol. II, p. 184.

Petrova, L.R. 1965. The morphology of the reproductive organs of *Melocanna bambusoides*. *Botanischesky Zhurnal* **50**: 1288–1304.

Philip, V.J. 1972. On the embryogenesis in *Bambusa arundinacea* Willd. *Curr. Sci.* **40**: 151.

Philip, V.J. and B. Haccius. 1976. On the embryogenesis in *Bambusa arundinacea* Willd. and Structure of mature embryo. *Beitr. Biol. Pfl.* **52:** 83–100.

Philipson, M.N. 1977. Haustorial synergids in *Cortaderia* (Gramineae). *New Zel. J. Bot.* **15**: 777–778.

Philipson, M.N. 1978a. Apomixis in *Cortaderia jubata* (Gramineae). *New Zel. J. Bot.* **16**: 45–60.

Philipson, M.N. 1978b. Nucellar degeneration in *Cortaderia* (Gramineae). *Protoplasma* **95**: 361–370.

Philipson, M.N. 1981. The haustorial synergids of *Cortaderia* (Gramineae) at maturity. *Acta Soc. Bot. Pol.* **50:** 151–160.

Philipson, M.N. and H.E. Connor. 1984. Haustorial synergids in Danthoid grasses. *Bot. Gaz.* **145**: 78–82.

Pi, P. and C. Chao. 1974. Microsporogenesis in *Paspalum longifolium* and *P. commersonii* on two different levels. *Cytologia*

Pitman, M.W., B.L. Burson and E.C. Bashaw. 1987. Phylogenetic relationships among *Paspalum* species with different base chromosome numbers. *Bot. Gaz.* **148**: 130–135.

Poddubnaja Arnoldi, V.A. 1964. *Embryology of Angiosperms* (in Russian). Moscow.

Poddubnaja–Arnoldi, V.A. and I.A. Ivanow. 1971. Macrosporogenese et development dn gametophyte femelle chez certains hybrides eloignes de la famille des Graminae. *Ann. Univ. Reims* **9**:

Poddubnaja–Arnoldi, V.A. and M.A. Makhaline. 1969. Microsporogenese et Spermiogenese Chez certains hybrids eloignes de la famille des Graminae. *Rev. Cytol. et Biol. Veget.* **32:** 59–66.

Pomeranz, Y. and D.B. Bechtel. 1978. Structure of cereal grains as related to end–use properties. In: Hultin, H.O. & M. Millner (eds.) *Post–harvest biology and biochemistry.* Food and Nutrition Press, Westport, USA.

Poverene, M.M. and P.W. Voigt. 1995. Identification of apomictic and sexual *Eragrostis curvula* (Schrad.) Nees hybrids by Isozyme analysis. *Mendeliana* **11**: 29–36.

Pritchard, A.J. 1967. Apomixis in *Brachiaria decumbens. J. Aust. Inst. Agric. Sci.* **33**: 264–265.

Pritchard, A.N. 1970. Melosis and embryo sac development in *Urochloa mosambieansis* and three *Paspalum* species. *Aust. J. Agric. Res.* **21**: 649–652.

Pupili, F., M.E. Caceres, C.L. Quarin and S. Arcioni 1997. Segregation analysis of RFLP markers reveals a tetrasomic inheritance in apomictic *Paspalum simplex. Genome* **40:** 822–828.

Quarin, C.L. 1986. Seasonal changes in the incidence of apomixis of diploid, triploid and tetraploid plants of *Paspalum cromyorrhizon. Euphytica* **35:** 515–522.

Quarin, C.L. 1992. The narure of apomixis and its origin in Panicoid grasses. *Apomixis News Letter* **5:** 8–15.

Quarin, C.L. 1994. A tetraploid cytotype of *Paspalum durifolium*: cytology, reproductive behaviour and its relationship to diploid *P. intermedium. Hereditas* **121:** 115–118.

Quarin, C.L. 1999. Effect of pollen source and pollen ploidy on endosperm formation and seed set in pseudogamous apomictic *Paspalum notatum. Sex. Plant Reprod.* **11:** 331–335.

Quarin, C.L. and B.L. Burson. 1991. Cytology of sexual and apomictic *Paspalum sp. Cytologia* **56:** 223–228.

Quarin, C.L. and I. Caponio. 1995. Cytogenetics and reproduction of *Paspalum dasypleureum* and its hybrids with *P. urvillei* and *P. dilatatum* ssp. *flavescens. Int. J. Plant Sci.* **156:** 233–235.

Quarin, C.L. and W.W. Hanna. 1980a. Chromosome behaviour, embryo sac development and fertility of *Paspalum modestum, P. boscianum* and *P. conspersum. J. Heredity* **71**: 419–422.

Quarin, C.L. and W.W. Hanna. 1980b. Effect of three ploidy levels on meiosis and mode of reproduction in *Paspalum hexastachyum. Crop Sci.* **20**: 69–75.

Quarin, C.L., W.W. Hanna and A. Fernandez. 1982. Genetic studies in diploid and tetraploid *Paspalum* species. *J. Hered.* **73**: 254–256.

Quarin, C.L. and G.A. Norrmann. 1987. Relaciones entre elnumero cromosomas, su compartimento en la meiosis y sistema reproduction del genero *Paspalum. Anales del IV. Congreso Latinamericano de Botanica* (Bogota) **3:** 25–34.

Quarin, C.L. and G.A. Norrmann. 1987. Cytology and reproductive behaviour of *Paspalum equitants, P. ionanthum* and their hybrids with diploid and tetraploid cytotypes of *P. cromyorrhizon. Bot. Gaz.* **148:** 386–391.

Quarin, C.L. and G.A. Norrmann. 1990. Interspecific hybrids between five *Paspalum* species *Bot. Gaz.* **151**: 366–390.

Quarin, C.L., G.A.Norrmann and F.Espinosa. 1998. Evidence for autoploidy in apomictic *Paspalum rufum. Hereditas* **129:** 119–124.

Quarin, C.L., M.T. Pozzobon and J.F.M. Valls. 1996. Cytology and reproductive behaviour of diploid, tetraploid and hexaploid germplasm accessions of a wild forage grass: *Paspalum compressifolium. Euphytica* **90:** 345–349.

Quarin, C.L. and M.H. Urbani. 1990. *Apomixis News Letter* **2:** 44.

Quarin, C.L., J.F.M. Valls and M.H. Urbani. 1997. Cytological and reproductive behaviour of *Paspalum atratum*, a promising forage grass for the tropics. *Tropical Grasslands* **31:** 114–116.

Rabau, T., B. Longly et B.P. Lovant. 1986. Ontogenese des sacs embryonnaires non reduits chez *Eragrostia curvuls. Can. J. Bot.* **64**: 1778–1785.

Raj, A.Y. 1969. Histological studies in male sterile and male fertile *Sorghum. Indian J. Genet. Plant Breed.* **28**: 335–345.

Raju, P.S.G. 1980. *Embryological and Histochemical studies of some crop plants (Gramineae).* Ph.D. Thesis, Nagpur Univ.

Raju, P.S.G. and P.K. Deshpande. 1981. The development in *Triticum durvm. Proc. 68th Indian Sci. Congr. Part III. Abst. pp* 48.

Ramaiah, K., N. Parthasarathi and S. Ramanujam. 1933. Haploid plant in rice (*Oryza sativa*). *Curr. Sci.* **1**: 277–278.

Ramaiah, K., N. Parthasarathy and S. Ramanujam. 1935. Polyembryony in rice *Oryza sativa*. *Indian J. Agric. Sci.* **5**: 119–124.

Ramu, J. S.L. Hatch, M.A. Hussey and E.C. Bashaw 1996. Morphology of *Pennisetum orientale* (Poaceae: Paniceae). *SIDA Contr. Bot.* 17: 163–171.

Randolph, L.F. 1936. Developmental morphology of the caryopsis in maize. *J. Agric. Res.* **53**: 881–916.

Randolph, L.F. and H.E. Fischer. 1939. The occurrence of parthenogenetic diploids in tetraploid maize. *Proc. Nat. Acad. Sci. Wash.* **25**: 161–164.

Rangaswamy, K. 1935. On the cytology of *Pennisetum typhoideum*. *J. Indian Bot. Soc.* **14**: 125–131.

Rao, N.G.P. and L.L. Narayana. 1968. Apomixis in grain *Sorghum*. *Indian J. Genet.* **28**: 121–127.

Rao, N.G.P. and U.R. Murthy. 1972. Further studies on obligate apomixis in grain *Sorghum* L. *Indian J. Genet.* **32**: 379–383.

Rao, N.G.P., L.L. Narayana and R.N. Reddy. 1978. Apomixis and its utilisation in grain *Sorghum* 1. Embryology of two apomictic parents. *Caryologia* **31**: 427–433.

Rao, P.N. 1974. Male parthenogenesis in tetraploid job's tears. *Heredity* **32**: 412–414.

Rao, P.N. and D.S. Narayana. 1980. Occurrence and identification of semigamy in *Coix aquatica* (tribe Maydeae). *J. Hered.* **71**: 117–120.

Ravi, S.B. 1993. Apomixis in Sorghum line R 473 – a critical analysis of published work. *Curr. Sci.* **64:** 306–314.

Reddy, P.S. 1977. Evolution of the apomictic mechanism in the Gramineae. A concept. *Phytomorphology.* **27**: 45–49.

Reddy, P.S. and R. D'Cruz. 1969a. Mechanism of apomixis in *Dicanthium annulatum*. *Bot. Gaz.* **130:** 71–79.

Reddy, P.S. and R. D'Cruz. 1969b. Polyembryony in *Dicanthium annulatum*. *Bot. Gaz.* **130**: 162–165.

Reddy, P.S. and R. D'Cruz. 1969c. Development of caryopsis in *Dicanthium armatum* (Hook. f..) Blatt. et McCann. *J. Indian Bot. Soc.* **48**: 351–359.

Reddy, R.N., L.L. Narayana and N.G.P. Rao. 1979. Apomixis and its utilisation in grain *Sorghum* 2. Embryology of F_1 progeny of reciprocal crosses between R 473 and 302. *Proc. Indian Acad. Sci.* B88: 455–461.

Reddy, G.B. and M.V. Reddi. 1974. Cytohistological studies on certain male sterile lines of Pearl millet (*Pennisetum typhoides*). *Cytologia* **39**: 585–589.

Reeder, J.R. 1953. The embryo of *Streptochaeta* and its bearing on the homology of the coleoptile. *Amer. J. Bot.* **40**: 77–80.

Reeder, J.R. 1956. The embryo of *Jouvea pilosa* as further evidence for the foliar nature of the coleoptile. *Bull. Torrey Bot. Cl.* **83**: 1–4.

Reeder, J.R. 1957. The grass embryo in systematics. *Amer. J. Bot.* **44**: 756–768.

Reeder, J.R. 1962. The bambusoid embryo: A reappraisal. *Amer. J. Bot.* **49**: 639–641.

Reeder, J.R. 1982. Systematics of the tribe *Orcutlineae* (Gramineae) and the description of a new segregate genus, *Tuctoria. Amer. J. Bot.* **69**: 1982–1095.

Reeves, R.G. 1928. Partition wall formation in the pollen mother cells of *Zea mays. Amer. J. Bot.* **15**: 114–122.

Rendle, A.B. 1930. *The classification of flowering plants.* Cambridge Univ. Press, Cambridge (2nd ed. 1938).

Renner, O. 1916. Zur terminologie des Pflanzenlichen Generation–Swechsels. *Biol. Zentralblant.* **36**: 337–374.

Reusch, J.D.H. 1961. The relationship between reproductive factors and seed set in *Paspalum dilatatum. S. Afr. J. Agric. Sci.* **4:** 513–530.

Rocheva, G.P. and N.N. Taskaeva. 1981. *Referativnyi Zhurnal* **8**: 392.

Rodrigo, P. 1926. A case of polyembryony in rice. *Philipp. Agric.* **14**: 629–630.

Rogers, S.O. and R.S. Squatrano. 1983. Morphological staging of wheat caryopsis development. *Amer. J. Bot.* **70**: 308–311.

Romanov, I.E. 1972. Development du gametophyte male chez le Froment (*Triticum aestivum* L.) apres les observation *in vivo*. *Ann.* Univ. Reims. **9**.

Roques, D. and P. Feldmann. 1996. Development de l'anthere et du grain de pollen chez une espece sauvage apparentee a la canne a sucre (*Saccharum spontaneum*). *Canad. J. Bot.* **74**: 788–795.

Rossa, W.M. and J.A. Wilson. 1969. Polyembryony in *Sorghum. Crop. Sci.* **9**: 842–843.

Rost, T.L., P.I. Artucio and E.B. Risly. 1984. Transfer cells in the placental pad and caryopsis coat of *Pappophorum snbbulbosum* Arech. (Poaceae). *Amer. J. Bot.* **71**: 948–957.

Rost, T.L. and N.R. Lersten. 1970. Transfer aleurone cells in *Setaria leutescens* (Gramineae). *Protoplasma* **71**: 403–408.

Rowley, J.R. 1962a. Strandard arrangement of sporopollenin in the exine microspores of *Poa annua. Science* **137**: 526.

Rowley, J.R. 1962b. Nonhomogenous sporopollenin in microspores of *Poa annuua. Grana palynol.* **3**: 3–19.

Rowley, J.R. 1963. Ubisch body development in *Poa annua. Grana palyn.* **4**: 25–36.

Rowley, J.R. 1964. Formation of pore in pollen of *Poa annua.* In: Linskens, H.F. (ed.) *Pollen physiology and fertilization.* North Holland Publ., Amsterdam, The Netherlands. pp. 59–69.

Rowley, J.R., K. Muhlethaler and A. Frey–Wyssling. 1959. A route for the transfer of materials through the pollen grain. *J. Biophys. Biochem. Cytol.* **6**: 537–538.

Rowley, J.R. and J.J. Skvarla. 1974. Plasma membrane–glycocalyx origin of ubisch body wall. *Pollen Spores* **16**: 441–448.

Rudall, P.J. and S. Dransfied. 1989. Fruit structure and development in *Dinochloa* and *Ochlandra* (Gramineae, Bambusoidiae). *Ann. Bot.* (London) **63**: 29–38.

Rudall, P.J. and H.P. Linder. 1988. The embryo sac and nucellus in Restionaceae and Flagellariaceae. *Amer. J. Bot.* **75**: 1777–1786.

Russell, S.D. 1979. Fine structure of megagametophyte development in *Zea mays. Canad. J. Bot.* **57**: 1093–1110.

Russel, S.D. 1992. Double fertilization. In: *Sexual Reproduction in Flowering Plants.* S.D. Russel and C. Dumas (eds.) *Int. Rev. Cytol.* **140**: 357–388.

Rychlewski, J. 1958. Cytoembryological investigations on *Sesleria tatrae. Acta Biol. Cracov. Bot.* **1**: 103–113.

Rychlewski, J. 1961. Cyto–embryological studies in the apomiotic species *Nardus stricta. Acta Biol. Cracov. Bot.* **4:** 1–23.

Saha, B. 1956. Studies on the development of the embryo of *Oryza sativa* and the homologies of its parts. *Proc. Natl. Inst. Sci. India* Part B. **22**: 86–101.

Santos, J.K. 1933. Morphology of the flower and mature grain of Philippine rice. *Philipp. J. Sci.* **52**: 475–503.

Sapre, A.B. 1964. *Embryological studies in the genus Oryza.* Ph.D. Thesis, Nagpur University.

Sapre, A.B. 1968. Embryological studies in a triploid plant of rice. *J. Univ. Poona (India)* **34**: 85–89.

Sapre, A.B. 1970a. Embryological studies in haploid rice. *Sci. & Cult.* **36**: 225–226.

Sapre, A.B. 1970b. Embryological studies in relation to the sterility in the F_1 between *Oryza sativa* and *Oryza alta. Sci. & Cult.* **36**: 667.

Sapre, A.B. 1976. Embryology of multicaryoptic variety of rice. In: Indo–Soviet Symposium on Embryology of Crop Plants. Univ. of Delhi. p. 36.

Saran, S. and J.M.J. de Wet. 1969. A structural peculiarity observed in the sexual embryo sacs of *Dichanthium intermedicm. Can. J. Bot.* **47**: 1205–1206.

Saran, S. and J.M.J. de Wet. 1970. Environmental control of reproduction in *Dicanthium intermedium* (Gramineae). *Bull. Torrey Bot. Cl.* **97**: 6–13.

Saran, S. and I.N. Mishra. 1986. Apomixis in some *Pennisetum* spp. In: G.K. Manna & U. Sinha (eds.). *Perspectives in Cytology and Genetics*. p. 779–784.

Sass, J.E. 1946. The development of endosperm and antipodal tissue in Argentine waxy maize. *Amer. J. Bot.* **33**: 791–795.

Sass, J.E. and G.F. Sprague. 1950. The embryology of germless maize. *Iowa State Coll. J. Sci.* **24**: 209–218.

Satyamurthy, T.V. Ch. 1981. *A contribution in the Embryology of Poaceae.* Ph.D. Thesis, Andhra University, Visakhapatnam.

Satyamurty, T.V. Ch. 1983. Structure and development of the caryopsis in *Sporobolus coromandelianus. Curr. Sci.* **52**: 549–550.

Satyamurty, T.V. Ch. 1984. Development of the caryopsis in *Chlonachne koenigii. Proc. Indian Acad. Sci.* (Plant Sci.) **93**: 567–570.

Satyamurty. T.V. Ch. 1985a. Embryology of *Elytrophorus spicata. Phytomorophology* **35**: 11–15.

Satyamurty, T.V. Ch. 1985b. Life history of *Dinebra retroflexa. Acta Bot. Indica* **13**: 366–368.

Satyamurthy, T.V. and V. Seshavatharam. 1983. Apospory in *Eremopogen foveolatus. Curr. Sci.* **53**: 1158–1159.

Satyamurthy, T.V. Ch. and V. Seshavatharam. 1984. Sterility in *Phragmites communis. Curr. Sci.* **53:** 1158–1159.

Savchenko, M.I. 1960. Anomalies in the structure of angiosperm ovules. *Dokl. Akad. Nauk. SSR* **130**: 15–17.

Savchenko, M.I. 1973. *Ovule morphology in angiosperms.* Nauka, Leningrad.

Savidan, Y. 1975. Heredite de l'apomixie. Contribution a l'etude de l'heredite de l'apomixie sur *Panicum maximum* Jacq. (analyse des sacs embryonnaires) *Cah ORSTOM* Ser. *Biol.* **10**: 91–95.

Savidan, Y. 1978. L'apomixie gametophytique chez les Graminees et soa utilization en Ameliortaion des plantes. *Ann. Amelior Plant* (Paris) **28**: 1–9.

Savidan, Y. 1978. Genetic control of facultative apomixis and application in breeding *Panicum maximum.* XIV International Congr. Genetic. Moscow. 21–30/8.

Savidan, Y. 1980. Chromosomal and embryological analyses in sexual x apomictic hybrids of *Panicum maximum* Jacq. *Proc. XIV Intl. Grassland Congr. Lexington Ky.* pp. 182–184.

Savidan, Y. 1982a. Nature et heredite de l'apomixi chez *Panicum maximum* Jacq. *Trav. Doc.* ORSTOM **153**: 1–159.

Savidan, Y. 1982b. *Nature et heredite de l'apomixie chez Panicum maximum Jacq.* Ph.D. Thesis. Universite Paris, XI. France.

Savidan, Y. 1983. Embryological analysis of facultative apomixis in *Panicum maximum. Crop. Sci.* **22**: 467–469.

Savidan, Y.H. 1992. Progress in research on apomixis and its transfer to major grain grain crops. In *Reproductive Biology and Plant Breeding.* Y. Dattee, C. Dumas and A. Gallais (eds.) Springer–verlag. pp. 269–279.

Savidan, Y. and J. Berthaud. 1994. Maize x *Tripsacum* hybridization and the potential for apomixis transfer for maize improvement. *Biotechnology in Agriculture and Forestry* Vol. 25. Maize. Y.P.S. Bajaj (ed.). Springer–Verlag. Berlin p. 69–83.

Savidan, Y.H. and C.F. Crane (ed.). 1989, 1990. Apomixis Newsletter 1 & 2. ORSTOM & Dept. Soil. Crop. Sci. Texas A & M. Univ. Coll. Station, USA.

Savidan, Y. and M. Dujardin. 1992. Apomixie: la prochaine revolution verte? *La Recherche* **241**: 326–334.

Savidan, Y., O. Leblanc and J. Berthaud. 1994. Transferring apomixis to maize. In: *Apomixis: Exploiting hybrid vigor in rice.* G.S. Khush (ed.) IRRI, Manila. p. 23–29.

Savidan, Y. and J. Pernes. 1982. Diploid–tetraploid–dihaploid cycles and the evolution of *Panicum maximum* Jacq. Evolution **36**: 596–600.

Savidan, Y., J. Pernes and R. Chaume. 1979. Diploid–tetraploid–dihaploid cycles and their role in the organization of the variability and evolution of *Panicum maxicum.* Jacq. *Can. J. Genet. Cytol.*

Schnarf, K. 1929. *Embryologie der Angiospermen.* Berlin.

Schnarf, K. 1931. *Vertgleichende Embryologie der Angiospermen.* Berlin.

Schnarf, K. 1933. Bedeutung des embryologischen forschung fur dad naturliche system der Pflanzen. *Biol. Gen.* **10**: 271–288.

Schnarf, K. 1937. Ziele und Wege der vergleichenden embryologie der Blutenflanzen. *Verh. Zool. Bot. Fges. Wien.* **86–87**: 140–147.

Schwab, C.A. 1971. Callose in megasporogenesis of *Diarrhena* (Gramineae). *Can. J. Bot.* **49:** 1523–1524.

Schwartz, O. 1930. Pontederiaceae. In *Engler and Prantl. Die naturlichen Pflanzenfamilien.* ed. **2.**15: 181–188.

Seshavatharam, V. 1977. The mechanism of Apomixis in some forage grasses. In: *Recent Trends and contacts between cytogenetics, Embryology and Morphology* (Ed. V.R. Dnyanagar *et al.*) Today & Tomorrow, New Delhi, pp. 323–330.

Seshavatharam, V. 1983. *Embryological studies on Apomixis in some forage grasses.* ICAR Project Report. Mss. Andhra Univ. Waltair.

Seshavatharam, V. and S.V. Bhasakara Rao. 1983. A contribution to the life history of *Sporobolus tremulus.* 6th All India Bot. Conf. Abstr. p. 49.

Seshavatharam, V. and S.V. Bhasakara Rao. 1984. A contribution to the life history of *Eremopogon foveolatus. Proc. 71st Indian Sci. Congr. Part III.* Abst. 116.

Seshavatharam, V. and T.V. Ch. Satyamurty. 1976. Embryology of Forage grasses. *Abs. Proc. Indo–Soviet Symp. on Embryology of Crop Plants,* Delhi.

Seshavatharam, V. and T.V. Ch. Satyamurty. 1978. Protogyny in *Elytrophorus spicata* (Willd.) A. Camus. *Curr. Sci.* **47:** 90.

Seshavatharam, V. and T.V. Ch. Satyamurty. 1982a. Embryology of *Sporobolus coromandelianus* and *Diplachne fusca. Proc. 69th Indian Sci. Congr. Part III.* Abst. p. 94.

Seshavatharam, V. and T.V. Ch. Satyamurty. 1982b. On the occurrence of Hypostase in Grasses. *Proc. 69th Indian Sci. Congr. Part III.* Abst. pp. 94–95.

Shadowsky, A.E. 1962. Der antipodiale Apparat bei Gramineen, *Flora* **120**: 344–370.

Shamakumari, K. 1960. Cytogenetic investigations in Paniceae. Occurrence of apospory in a diploid species of *Panicum, P. antidotale. Curr. Sci.* **29**: 191.

Shanthamma, C. 1973. *Studies in Poaceae.* Ph.D. Thesis, Mysore Univ., Mysore.

Shanthamma, C. 1979. Reproductive behaviour of *Pennisetum macrostachyum* and a new basic Chromosome number in the genus *Pennisetum. Bull. Torrey Bot. Cl.* **106:** 73–78.

Shanthamma, C. 1982. Apomixis in *Cenchrus glaucus* Mudaliar et Sundarraj. *Proc. Indian Acad. Sci. (Plant Sci.)* **91**: 25–36.

Shanthamma, C. and K.N. Narayan. 1976–77. Studies in Poaceae (Gramineae). *Journ. Mysore Univ.* Sect. B. **27**: 302–305.

Shanthamma, C. and K.N. Narayan. 1977. Formation of nucellar embryos with total absence of embryo sacs in two species of Gramineae. *Ann. Bot.* **41**: 469–470.

Sharma, M.L., R.K. Bhanwra and S. Kaur. 1981. Seed sterility in *Dactyloctenium sindicum. Curr. Sci.* **50**: 771–772.

Sharman, B.C. 1942. A twin seedling in *Zea mays* L. Twinning in Gramineae. *New Phytol.* **41**: 125–129.

Sherman, R.A., P.W. Voigt, B.L. Burson and C.L. Dewald. 1991. Apomixis in diploid x triploid *Tripsacum dactyloides* hybrids. *Genome* **34**: 528–532.

Sherwood, R.T. 1995. Nuclear DNA amount during sporogenesis and gametogenesis in sexual and aposporous buffel grass. *Sexual Plant Reproduction.* **8**: 85–90.

Sherwood, R.T., C.C. Berg and B.A. Young. 1994. Inheritance of apospory in buffel grass. *Crop Sci.* **34**: 1490–1494.

Sherwood, R.T. and D.L. Gustine. 1994. Modifying embryo sac development and strategies for cloning the apomixis gene from buffelgrass. In: *Apomixis: Exploiting hybrid vigor in rice.* G.S. Khush (ed.) IRRI, Manila p. 47–50.

Sherwood, R.T., B.A. Young and E.C. Bashaw. 1980. Facultative apomixis in Buffel grass. *Crop. Sci.* **20**: 375–379.

Shishkinskaya, N.A. and A.V. Borodko. 1987. Apomixis in *Festuca drymeja. Biol. Nauki* (Moscow). **1**: 84–89.

Shehata, A. 1995. Embryological studies in *Zingeria trichopoda. Phytomorphology.* **45**: 185–190.

Shobha, J. 1988. *Cytoembryological studies in a few species of Panicoideae.* Ph.D. Thesis, Unive. of Mysore.

Shobha, P. and A.N. Sindhe. 1987. Synergid embryo in *Pennisetum pedicellatum. Proc. 74th Indian Sci. Congr. Part III.* Abstracts p. 154.

Shobha, J. and A.N. Sindhe. 1990. Nucellar embryos in *Pennisetum pedicellatum* Trin. *Proc. XI Intern. Symp. on Embryology and seed reproduction.* Leningrad.

Shobha, J. and A.N. Sindhe. 1987. Life history of *Jansenella griffithiana. Acta Bot. Indica.* **15**: 316–318.

Shobha, J. and A.N. Sindhe. 1992. Embryology of *Acroceras munroanum* (Balansa) Henr. (Poaceae). *J. Mysore Univ.* Sect. B. **32**: 523–531.

Shobha, J. and A.N. Sindhe. 1995–96. Embryology of an endangered aquatic grass *Limnopoa meeboldii* (Fischer) C.E. Hubb. *J. Mysore Univ.* **34**: 1–11.

Sidhu, B.S.S. 1993. Combining ability analysis for reproductive traits in guinea grass. *Tropical Agriculture* **70:** 252–255.

Simpson, C.E. and E.C. Bashaw. 1969. Cytology and reproductive characteristics in *Pennisetum setaceum. Amer. J. Bot.* **56**: 31–36.

Simpson, C.E. and E.C. Bashaw. 1969. Cytology and reproductive characteristics in *Pennisetum setaceum*. *Amer. J. Bot.* **56**: 31.

Sindhe, A.N. 1976a. *Reproductive behaviour of Pennisetum squammulatum. Proc. Indo Soviet Symp. on Embryology of crop plants.* Abstr. p. 41–42.

Sindhe, A.N. 1976b. Supernumerary chromosomes in *Pennisetum squamulatum* Fresen. *Curr. Sci.* **45**: 526.

Sindhe, A.N. and J. Shobha. 1990. Origin shoot apex in *Pennisetum pedicellatum.* Trin. *XI International Symp. on Embyrology and seed reproduction.* Leningrad. p. 159.

Sindhe, A.N.R., B.G.L. Swamy and G.D. Arekal. 1980. Synergid embryo in *Pennisetum squamulosum. Curr. Sci.* **49**: 914–915.

Singh, S.P. and H.H. Hadly. 1961. Pollen abortion and cytoplasmic male sterility in *Sorghum. Crop. Sci.* **1:** 410–432.

Skalinska, M. 1952. Embryological studies in *Poa alpina* var. *vivipara. Bull. Internat. Acad. Pol.* B: 253–283.

Skalinska, M. 1959. Embryological studies in *Poa granitica,* an apomictic species of the Carpathian range. *Acta Biol. Cracov. Bot.* **2**: 91–112.

Skovsted, A. 1939. Cytological studies in twin plants. *Compte Rend. Lab. Carlsberg (Copenhague). Ser. Phys.* **22**: 427–446.

Skvarla, J.J. and D.A. Larson. 1966. Fine structural studies of *Zea mays* pollen. I. Cell membranes and exine ontogeny. *Amer. J. Bot.* **53**:1112–1125.

Smith, B.W. 1948. Hybridity and Apomixis in the perennial grass *Paspalum dilatatum. Genetics* **33**: 628–629.

Smith, L.B. 1934. Geographical evidence on the lines of evolution in the Bromeliaceae. *Bot. Jahrb.* **66**: 446–468.

Smith, R.L. 1972. Sexual reproduction in *Panicum maximum. Crop. Sci.* **12**: 624–627.

Snell, R.S. 1936. Anatomy of the spikelet and flowers of *Carex, Kobresia* and *Uncinis. Bull. Torrey Bot. Cl.* **63**: 277–295.

Snyder, L.A. 1957. Apomixis in *Paspalum secans. Amer. J. Bot.* **44**: 318–324.

Snyder, L.A., A.R. Hernandez and H.E. Warmke. 1955. The mechanism of Apomixis in *Pennisetum ciliare. Bot. Gaz.* **116:** 209–221.

Sokolova, V.A., B. Kindiger, C. Dewald and I.V. Khatyshova. 1996a. A comparative analysis of apomictic reproduction in maize–*Tripsacum* hybrids in gamma grass. *Dokl. Akad. Nauk.* **346**: 845–847.

Sokolova, V.A., L.A. Lukina and I.V. Khatypov. 1996b. Perspectives of developing apomixis in maize *Dokl. Akad. Nauk.* **347**: 714–717.

Solntzeva, M.P. 1974. Disturbances in the process of fertilization in angiosperms under hemigamy. In: H.F. Linskens (ed.) *Fertilization in Higher Plants.* North Holland Publ. Co. Amsterdam. pp. 311–324.

Soderstrom, T.R., K.W. Hilu, C.S. Campbell and Barkworth. (ed.) 1986. *Grass Systematics and Evolution.* Smithsonian Inst. Press. Washington, D.C.

Sonpipare, S.A. 1984. Studies in the Gramineae: Polyembryony in *Eragrostiella bifaria. Proc. 71st Indian Sci. Congr. Part III.* Abst. p. 258.

Sonpipare, S.A. and M.D. Padhye. 1980. Studies in the Gramineae. Male and female gametophyes of *Alloteropsis cimicina. Proc. 67th Indian Sci. Congr. Part III.* Abst. p. 45.

Souciet, J.L. 1978. *Controle de l'inhibition de la germination chez une espece apomictiquce, le Panicum maximum.* These 2385, Univ. Paris–Sud.

Souegcs, R. 1924. Embryogenie des Graminees. Development de l'embryon chezle *Poa annua* L. *C.R. Acad. Sci. Paris* **178:** 860–862.

Spies, J.J. K.C. Klopper and B. Visser. 1999. Apomixis in the genus *Pentaschistis* (Arundinoideae). *Bothalia* **29**. (in press).

Sprague, G.F. 1932. The nature and extent of heterofertilization in Maize. *Genetics* **17**: 358–368.

Sreenivasa Rao, K. 1997. *Embryological investigations in Poaceae*. Ph.D. Thesis, Andhra University.

Sree Rangaswamy, S.R. and L.D. Vijendra Das. 1973. Embryology of *Oryza sativa* Linn. diploid and autotetraploid forms. *Proc. Indian Acad. Sci.* **B 77:** 234–242.

Srivastava, A.K. 1978. Agamospermy in some natural hybrids of *Bothriochloa* and *Dichanthium. First All India Bot. Conf.* Abstract p. 45.

Srivastava, A.K. 1982. Apomixis in *Paspalum paspaloides. Acta Bot. Indica* **10:** 111–113.

Stapf, O. 1904. On the fruit of *Melocanna bambusoides* an endospermless viviparous genus of Bambuseae. *Trans. Linn. Soc. London* (2) *Bot.* **6**: 401–425.

Stebbins, G.L. 1941. Apomixis in the Angiosperms. *Bot. Rev.* **7**: 507–542.

Stebbins, G.L. 1950. *Variation and evolution in plants.* Columbia Univ. Press, New York.

Steer, M.W. 1977. Differentation of the tapetum in *Avena.* I. The cell surface. *J. Cell Sci.* **25**: 125–138.

Stenar, H. 1932. Parthenogenesis in der Gattung *Calamagrostis,* embryobildung die *Calamagrostis obtusata* and *C. purpurea*. *Arkiv. f. Wiss. Bot.* **25:** 1–8.

Stover, E.L. 1937. The embryo sac of *Eragrostis cilianensis* — A new type of embryo sac and a summary of grass embryo sac investigations. *Ohio J. Sci.* **37:** 172–181.

Stratton, M.E. 1923. The embryology of the double kernel in *Zea mays* var. *polysperma. N.Y. Cornell Agri. Expt. Sta. Mem.* **69**:

Streetman, L.J. 1963. Reproduction of the love grasses, the genus *Erogrostis*–I, *E. chloromelae, E. curvula, E. lehmaniana* and *E. superba. Wrightia* **3**: 41–60.

Strydom, A. and J.J. Spies. 1994. Embryo sac development in some representatives of the tribe Cynodonteae (Poaceae) *Bothalia* **24**: 101–105.

Subramanyam, K. and H.S. Narayana. 1972. Some aspects of the floral morphology and embryology of *Flagellaria indica* Linn. *Advances Pl. Morphol.* pp. 211–217.

Suessenguth, K. 1919. Cited by Schnarf, K. 1929. In: *Embryologie der Angiospermen.* Berlin.

Suomalainen, E., A. Saura and J. Lokki. 1987. *Cytology and evolution in parthenogenesis.* CRC Press, Boca Raton.

Swamy, B.G.L. 1944. A reinvestigation of the embryo sacs of *Eragrostis cilianensis. Curr. Sci.* **13:** 103–104.

Szkukalek, A., M. Hausksecht and O. Erdelska. 1989. Some aspects of the developmental correlation of the maize embryo and endosperm. In: J. Pare and M. Bugnicourt (eds.) *Some aspects and actual orientations in Plant embryology.* Dedicatory volume of Prof. A. Lebegue. Univ. Press. Picardie, Amiens. pp. 245–250.

Takhtajan, A. 1969. *Flowering plants. Origin and dispersal.* Edinburgh.

Takhtajan, A. 1980. Outline of the classification of flowering plants (Magnoliophyta). *Bot. Rev.* **46**: 225–359.

Taliaferro, C.M. and E.C. Bashaw. 1966. Inheritance and control of obligate apomixis in breeding buffel grass, *Pennistetum ciliare. Crop Sci.* **6**: 473–476.

Tang, C.Y., K.F. Shertz and E.C. Bashaw. 1980. Apomixis in *Sorghum* lines and their F_1 progenies. *Bot. Gaz.* **141**: 294–299.

Tantravahi, R.V. 1965. Apomixis in teosinte. *M. G. C. Newsletter.* **39**: 70.

Tateoka, T. 1964. Notes on some grasses. I. Embryo structure of the genus *Oryza* in relation to systematics. *Amer. J. Bot.* **51**: 539–543.

Tateoka, T. 1968. Notes on *Calamagrostis hakonensis* (Gramineae). *Bull. Nat. Sci. Mus.* Tokyo **11**: 293–298.

Tateoka, T. 1969. Notes on some grasses XX. Systematic significance of the vascular bundle system in the mesocotyl. *Bot. Mag.* (Tokyo) **82**: 387–391.

Tateoka, T. 1973. A taxonomic study of the genus *Calamagrostis* On Mount Yakeishidake. *Bot. Mag.* (Tokyo) **86**: 103–120.

Tateoka, T. 1976. Chromosome numbers of the genus *Calamagrostis* in Japan. *Bot. Mag.* (Tokyo) **89**: 99–114.

Tateoka, T., A. Hiraoka and T.N. Tateoka. 1977. Natural hybridization in Japanese *Calamogrostis.* II *Bot. Mag.* (Tokyo) **90**: 193–209.

Tateoka, Y., K. Ogawa, T. Kawai and T. Wada. 1989. Scanning electron microscopic observations on morphogenesis of the panicle and spikelet in rice plants. *Jap. J. Crop Sci.* **58**: 119–125.

Terada, S. 1928. Embryological studies in *Oryza sativa. J. Coll. Agric. Hokkaido Imp. Univ. Sapporo, Japan* **19:** 245–260.

Terrel, E.E. 1971. Survey of occurrence of liquid and soft endosperm in grass genera. *Bull. Torrey Bot. Cl.* **98**: 264–268.

Terzijski, D. and M.A. Khristov. 1973. Cyto–embryological study of *Festuca pratensis* Huds., *Alopecuros pratensis* L., *Holcus lanatus* L. and *Agrostis vulgaris*. *Genet. Sci.* **6**: 45–57.

Thompson, W.P. and D. Johnson. 1945. The cause of incompatibility between barley and rye. *Can. J. Res.* **23C**: 1–5.

Thorne, R. 1968. Synopsis of putatively phylogenetic classification of the flowering plants. *Aliso* **6**: 57–66.

Tieghem, P. Van. 1897. Morphologie de l' embryon et de la plantule chez les Graminees et les Cyperaceae. *Ann. Sci. Nat. Bot. VIII.* **3**: 259–309.

Tillich, H.J. 1977. Vergleichende morphologische untessuchungen zur identitat der Gramineen – Primarwurzel. *Flora* **166**: 415–421.

Tinney, F.W. 1940. Cytology of parthenogenesis in *Poa pratensis*. *J. Agric. Res.* **60**: 351–360.

Torabinejad, J. and R.J. Mueller. 1993. Genome analysis of intergeneric hybrids of apomictic and sexual Australian *Elymus* species with wheat, barley and rye; implications for the transfer of apomixis to cereals. *Theor. Appl. Genet.* **86**: 288–294.

Tothill, J.C. 1968. Variation and apomixis in *Heteropogon contortus*, Gramineae. *Bot. Soc. Argent.* **12**: 188–201.

Tothill, J.C. and R.B. Knox. 1968. Reproduction in *Heteropogon contortus*. Gramineae. *Bol. Soc. Argent. Bot.* **12**: 188–201.

Tschermak–Woess, E. and U. Enzenberg–Kunz. 1965. Die Struktur der hoch endopolyploid, en Kerne im Endosperm von *Zea mays*, das anffallende Verhalten ihrer Nukleolen und ihr Endopolyploidiegrad *Planta* **64**: 149–169.

Tsvetova, M.I. and A.G. Ishin. 1996. Sorghum polyembryogenesis investigation for the purpose of receiving haploids for practical breeding *Selskhozyaistvennaya Biologiya*. pp. 86–91.

Tzvelev, N.N. 1989. The systems of Grasses (Poaceae) and their evolution. *Bot. Rev.* **55**: 141–203.

Untawale, A.G., P.K. Deshpande and K.B. Sharma. 1969. Studies on the Gramineae — I. Male and female gametophytes of *Eragrostis unioloides* (Retz.) Nees ex Steudel. *J. Indian. Bot. Soc.* **48**: 386–392.

Ustinova, E.L. 1960a. Specific structural characteristics of the female gametophyte and the phenomenon of polyembryony in corn *Zea mays*. *Bot. Zhur.* **45**: 764–767.

Ustinova, E.L. 1960b. Embryological study of female flowers from corn under various conditions of the life cycle. *Nauch. Dokl. Vysshei Shkoly Biol. Nauk.* **1**: 94–98.

Ustinova, E.L. 1960c. Cytoembryological investigations of the embryo sac and the process of fertilization in maize. *Zhur. Obschehei Biol.* **21**: 261–269.

Van Lammeren, A.A.M. 1981. Early events during embryogenesis in *Zea mays* L. *Acta Soc. Bot. Pol.* **50**: 289–290.

Vasil, I.K. 1960. Pollen germination in some Gramineae *Pennisetum typhoideum. Nature (London)* **187**: 1134–1135.

Venkateswarlu, J. and P.I. Devi. 1964. Embryology of some Indian grasses. *Curr. Sci.* **33**: 104–106.

Venkateswarlu, J. and M.K. Rao. 1966. Further studies on apomixis in *Coix aquatica. Maize Newslett.* **40**: 166.

Venkateswarlu, J. and P.N. Rao. 1975a. Apomictic maternal diploids in tetraploid job's tears. *Theor. App. Genet.* **45**: 274–276.

Venkateswarlu, J. and P.N. Rao. 1975b. Apomixis in *Coix aquatica* Roxb. *Ann. Bot.* **39**: 1131–1136.

Verboom, G.A., H.P. Linder and N.P. Barker. 1994. Haustorial synergids: An important character in the systematics of Danthoid grasses. *Amer. J. Bot.* **81**: 1601–1610.

Ville, J–Ph., B.L. Burson, E.C. Bashaw and M.A. Hussey. 1995. Early fertilization events in the sexual and aposporous egg apparatus of *Pennisetum ciliare* (L.) Link. *The plant Journal* **8:** 309–316.

Vielle Calzada, J.P., C.F. Crane and D.M. Stelly. 1996. Apomixis. The asexual revolution. *Science* **274:** 1322–1323.

Vijayaraghavan, C. and V.P. Rao. 1936. False polyembryony in *Setaria italica. Curr. Sci.* **4:** 820.

Vijendra Das, L.D. 1969. A note on twin megaspores and abnormal embryo sac in *Saccharum. Curr. Sci.* **38**: 248–249.

Vijendra Das, L.D. and S.R. Rangaswamy. 1974. Embryological studies in the sterile interspecific hybrid of rice between *O. sativa* (autotetraploid) and *O. perennis* (diploid). *Sci. Cult.* **40**: 253–254.

Visser, N.C. and J.J. Spies. 1994. Cytogenetic studies in the genus *Tribolium* (Poaceae: Danthonieae): II A report on embryo sac development, with special reference to the occurrence of apomixis in diploid specimens. *South African Journ. Bot.* **60**: 22–26.

Visser, N.C., J.J. Spies and H.J.J. Venter. 1999. Apomictic embryo sac development in *Cenchrus ciliaris* (Panicoideae). *Bothalia* (in press).

Vogl, E. 1947. Untersuchuntger uber die Tielunsrichtungen in pollen mutter zellen und bel due Blangalgae Chroococus. *Ost. Bot. Z.* **94**: 1–28.

Voigt, P.W. 1971. Discovery of sexuality in *Eragrostis curvala* (Schrad.) Nees. *Crop Sci.* **13**: 424–425.

Voigt, P.W. and E.C. Bashaw. 1972. Apomixis and sexuality in *Eragrostis curvula. Crop. Sci.* **12**: 842–847.

Voigt, P.W. and E.C. Bashaw. 1976. Facultative apomixis in *Eragrostis curvula. Crop. Sci..* **16**: 803–806.

Voigt, P.W. and B.L. Burson. 1983. Breeding of apomictic *Eragrostis curvula.* In *Proc. of the XIV International Grassland Congress.* (ed.) J.A. Smith and V.W. Hays. Boulder, USA, pp. 160–163.

Voigt, P.W., B.L. Burson and R.A. Sherman. 1992. Mode of reproduction in cytotypes of lehmann lovegrass. *Crop Sci.* **32**: 118–121.

Vollbrecht, E. and H. Sarah. 1995. Deficiency analysis of female gametogenesis in maize. *Developmental Genetics* **16**: 44–63.

Vorster, T.B. and H. Liebenberg. 1977. Cytogenetic studies in the *Eragrostis curvula* complex. *Bothalia* **12:** 215–221.

Vorster, T.B. and H. Liebenberg. 1984. Classification of embryo sacs in *Eragrostis curvula* complex. *Bothalia* **15**: 167–174.

Wakakuwa, S. 1934. Embryological studies on different seed development in reciprocal interspecific crosses of wheat. *Jap. J. Bot.* **7:** 151–185.

Walden, D.B. 1967. Male gametophyte *Zea mays*. Some factors influencing fertilization. *Crop Sci.* **7**: 441–444.

Walter, R. 1977. *Holcus mollis* L. in Poland. II. Studies in the origin of polyploid types. *Acta Biol. Cracov. Ser. Bot.* **20**: 113–131.

Wang, R.R.C., Z.W. Liu and J.G. Carman. 1995. The introduction and expression of apomixis in hybrids of wheat and *Elymus rectisetus* p. 317–319. In: Z.S. Li and Z.Y. Xin (ed.) *Proceedings of the 8th International Wheat Genetics Symposium*. China Agr. Sci. and Techn. Press, Beijing.

Warming. 1904. *A Handbook of Systematic Botany*. London.

Warming. 1912. *Froplanterne Kjobenhaun*. London.

Warmke, H.E. 1952. Apomixis in grasses with special reference to *Panicum maximum. Proc. 6th International Grassland Congr.* 209–215.

Warmke, H.E. 1954. Apomixis in *Panicum maximum*. *Amer. J. Bot.* **41**: 5–11.

Warmke, H.E. and S.J. Lee. 1977. Mitochondrial degeneration in Texas cytoplasmic male–sterile corn anthers. *J. Hered.* **68**: 213–222.

Warmke, H.E. and M.A. Overman. 1972. Cytoplasmic male sterility in *Sorghum*. I. Callose behaviour in fertile and sterile anthers. *J. Hered.* **68**: 103–112.

Warmke, H.E. and I.J. Lele. 1978. Pollen abortion in T cytoplasmic male sterile corn (*Zea mays*): A suggestive mechanism. *Science* **200**: 561–562.

Weatherwax, P. 1916. Morphology of the flowers of *Zea mays*. *Bull. Torrey Bot. Cl.* **48:** 127–144.

Weatherwax, P. 1917. The development of spikelet of *Zea mays*. *Bull. Torrey Bot. Cl.* **44**: 482–491.

Weatherwax, P. 1919. Gametogenesis and fecundation in *Zea mays* as the basis of xenia and heredity in the endosperm. *Bull. Torrey Bot. Cl.* **46**: 73–90.

Weatherwax, P. 1926. Persistence of the antipodal tissue in the development of the seeds of maize. *Bull. Torrey Bot. Cl.* **61**: 211–215.

Weatherwax, P. 1930. The endosperm of *Zea* and *Coix*. *Amer. J. Bot.* **17**: 371–380.

Weatherwax, P. 1934. Flowering and seed production in *Amphicarpon floridanum*. *Bull. Torrey Bot. Cl.* **61**: 211–215.

Weatherwax, P. 1955. Structure and development of reproductive organs. In: *Corn and Corn improvement*. New York.

Webber, J.M. 1940. Polyembryony. *Bot. Rev.* **6**: 575–598.

Weimarck, G. 1967a. Apomixis and sexuality in *Hierochloe australis* and in Swedish *H. odorata* on different polyploid levels. *Bot. Notiser* **120**: 209–235.

Weimarck, G. 1967b. Apomixis in *Hierochloe monticola* (Gramineae). *Bot. Notiser* **120**: 448

Weimarck, G. 1970. Apomixis and sexuality in *Hierochloe alpina* (Gramineae) from Finland and Greenland and *H. monticola* from Greenland. *Bot. Notiser* **123**: 495–504.

Weimarck, G. 1977. *Publ. Cairo Univ. Herb.* Nos. 7 and 8.

Weir, G.E. and H.M. Dale. 1960. A developmental study of wild rice *Zizania aquatica. Can. J. Bot.* **38**: 719–739.

Westermaier, M. 1890. Zur Embryologie der Phanerogamen, insbesendere uber die segennannten. Antipoden. *Nov. Acta Ksl. Leop Carol. Disch. Acta. Nat.***17**.

Wilson, K.J. 1993. *Proceedings of the International workshop on Apomoxis in Rice.* Rockfeller Foundation, New York.

Woodland, P.S. 1964. The floral morphology and embryology of *Themeda australis* (R.Br.) Stapf. *Aust. J. Bot.* **12**: 157.

Wu, Bo–ji, Xie, Ming–Tang, Chen, Yi–Ping, Jiang, Hui and Guo, Xue–Xing. 1991. Cytologic and embryologic studies on apomicts discovered from male sterility line C 1001 in rice (*Oryza sativa* L.). *Sci. China Ser. B. Chem. Life Sci. Earth Sci.* **34**: 823–831.

Wu, Shu–Biao, Shang, Yong–Jin, Han, Xue–Mei, Wang, Jiang–Xue, Niu, Tian–Tang, Zhang, Fu–Yao, Wei, Yao–Ming, Meng, Cue–Gang, Yan, Xi, Mei and Zheng, Jing–BO. 1994. Embryological study on apomixis in a sorghum line SSA–1. *Acta Botanica Sinica* **36**: 833–837.

Wu, Shu–Hsuen ane C.K. Ts Ai. 1965. Cytological studies of the double fertilization in rice. *Acta Bot. Sin.* **13**: 110–121.

Wunderlich, R. 1954. Uber das Antheren tapetum mit besonderer Berucksichtigung seiner Kernzahl. *Osterr. Bot. Zeit.* **101**: 1–63.

Xi, Xiang–Yuan and de–Cai Cui. 1983. Relationship between pollen and embryo sac development in wheat *Triticum aestivum* L. *Bot. Gaz.* **144**: 191–200.

Xi, Xiang–Yuan and de Mason. D.A. 1984. Relationship between male and female gametophyte development in Rye. *Amer. J. Bot.* **71**: 1067–1079.

Yakovlev, M.S. 1950. The structure of endosperm and embryo in cereals as a systematic feature: *Izv. Akad. Nauk. Arm. SSR Bot.* **1:** 121–128.

Yakovlev, M.S. 1970. On the embryology of *Melocanna bambusoides.* Ann. Bogor. **5**: 109–115.

Yakovlev, M.S. and O.P. Kamelina. 1973. Development of the embryo of *Melocanna bambusoides* Trin. *Bot. Zh.* **58**: 248–263.

Yamamoto, Y. 1936a. Uber das vorkommen von triploides pflanzen bei Mehrlings–keimlingen von *Triticum vulgare. Cytologia* **7:** 431–436.

Yamamoto, Y. 1936b. Ein haplo–diploides Zwillingspaar bei *Triticum vulgare* Vill. *Bot. Mag.* (Tokyo) **50**: 573–581.

Yamaura, A. 1933. Karyologischie and embryologische studies uber einige *Bambusa* Arten. *Bot. Mag.* Tokyo. **47**: 551–555.

Yamazaki, Y. 1937. Some notes on twin plants of common wheats. *Jap. Journ. Gen.* **13**: 193.

You, R. and W.A. Jensen. 1985. Ultrastructural observations on the mature megagametophyte and the fertilization in wheat. *Canad. J. Bot.* **63**: 163–178.

Young, B.A., R.T. Sherwood and E.C. Bashaw. 1979. Cleared pistil and thick sectioning techniciques for detecting aposporous apomixis in grasses. *Can. J. Bot.* **57**: 1668–1672.

Young, B.A., J. Schuez–Schaeffer and T.W. Carrol. 1979. Anther and pollen development in male sterile intermediate wheat grass plants derived from wheat x wheat grass hybrids. *Can. J. Bot.* **57**: 602–618.

Yoshida, O. 1963. Embryologische studien von *Arundinaria* Chino Malkino. *Coll. J. Arts Sci. Chiba Univ.* **4**: 33–41.

Yu Sh. and Ch. Y. Chao. 1979. Histochemical studies of ovary tissues during the embryo sac development in *Paspalum longifolium* Roxb. *Caryologia.* **32**: 147–160.

Zagorcheva, L.I. 1989. *Apomixis News Letter* **1**: 42.

Zhiping, Z., S. Rui–Yuan and T. Xi–Hua. 1990. Polyembryony in rice and its germination. *Chinese Journ. Bot.* **2**: 39–44.

Zimmerman, A. 1904. Ueber polyembryonie bei *Poa pratensis*. *Archiv des vereins der Freude der Naturgeschichte im Meklenberg*, Rostock. **58**: 107.

Zinn, J. 1904. Normal and abnormal germination of grass fruits. *Me. Agri. Exp. Sta. Bull.* 294.

SUBJECT INDEX

INDEX OF TAXA

AUTHOR INDEX